The Feel of Algorithms

The Feel of Algorithms

Minna Ruckenstein

 UNIVERSITY OF CALIFORNIA PRESS

University of California Press
Oakland, California

© 2023 by Minna Ruckenstein

Cataloging-in-Publication Data is on file at the Library of Congress.

ISBN 978-0-520-39454-4 (cloth)
ISBN 978-0-520-39455-1 (pbk.)
ISBN 978-0-520-39456-8 (ebook)

32 31 30 29 28 27 26 25 24 23
10 9 8 7 6 5 4 3 2 1

Contents

Preface

The story of this book began in September 2020 when I gave a public lecture at a local community college in Helsinki, Finland, where I tried to convince the audience that it is increasingly difficult to study everyday phenomena without recognizing their links to algorithmic technologies. With the constant growth in computational information-processing capacity and the spread of smartphones and sensors, digital data and algorithms are shaping the everyday; indeed, for growing masses of people, living without digital services and devices is unthinkable. Data gathering that feeds into algorithmic systems occurs when people pay with credit cards, use search engines, take part in customer loyalty programs, click advertisements, and upload content to social media platforms; it takes advantage of details such as the model of a person's smartphone, the time and duration of reading a news story online, and the history of goods and services purchased over the internet. Further examples that I gave covered pedometers and sleep trackers, detailing how the rules and procedures that they impose intertwine with everyday practices. I suggested that the culture we live in has become algorithmic in the sense that technologies organize our practices and interactions, promoting certain kinds of futures rather than others. What is critical, then, is that large data companies, mostly originating in the United States, have become controllers and gatekeepers that define how we navigate the online world, meaning that we are being managed by decisions made elsewhere, in corporations that have little interest in the society we want to inhabit.

After finishing my talk, I asked whether the audience had any questions. First there was a long silence, typical in Finland, where people tolerate silences much better than in many other parts of the world, thinking that it is better to say nothing than to share half-baked thoughts. Then a woman, perhaps in her sixties, raised her hand. She wanted to know how to improve online searches, as she had a hard time getting appropriate responses to her queries. I could not help feeling annoyed, as her question made it obvious that she had no clue what I had been talking about for the past hour. Her idea of the title of my talk—"Algorithmic Culture"—differed profoundly from what I had offered. Instead of a critical introduction focusing on the algorithmic logic penetrating the everyday, she had expected to get tips on how better to manage digital services like the Google search. I was complicating things rather than offering clarity and a way forward.

On my way home from the community college, I contemplated the annoyance that the woman's question had triggered in me and concluded that, ultimately, she had more reason for irritation, as she was the one who had been obliged to sit quietly and wait for my high-flying monologue to end. It is common to disregard the experiences of those who try to keep up with digital services in their everyday lives and favor those who use them comfortably, meanwhile speculating about humanity's destiny in the face of the paradigmatic changes promoted by sociotechnical relations.

Yet I thought how important it would be to bring these different perspectives together, under the same rubric of algorithmic culture, to offer a more comprehensive view of the experiential realms of the algorithmic perspective. After all, it is not only the professional elite who shape algorithmic culture, but also inexperienced users who feel the pressure to keep up with digital developments.

I had been studying how people's visions, ideas, aims, and behaviors are shaped by emerging technologies for years. I had ample resources to explore the experiential realms of the algorithmic; the collected material alone contained dozens of interviews conducted between 2017 and 2020. What I was lacking, however, was a framework for demonstrating that experiences with algorithms were patterned to such a degree that we could talk about algorithmic culture. Rather than placing experiences with algorithms into a ready-made framework of "algorithmic culture" or "algorithmic life," I wanted to investigate how algorithmic relations and imaginaries constitute culturally recognizable patterns and thereby shape the everyday. Thus, my aim was not to claim that algorithms have taken over our lives and are mercilessly controlling our minds and whole societies, but rather that the everyday is shaped and touched by algorithms and is coevolving with them. And to understand how this happens, we need to study how algorithms become participants in mundane visions and practices and what kinds of traces they leave in the process.

Thinking about the community college lecture, I decided that I would try to write about algorithmic culture in a manner that would be accessible to a larger public than my usual academic writings. One way to do this would be to review what I had been doing for nearly a decade and plot the evolution of the idea that algorithmic experiences are culturally patterned yet can tell a generalizable story of attempts to live well with algorithmic systems. Once I ventured on this journey, I was pushed in a direction that required that the feelings triggered by algorithms be taken seriously. This is what happens with open-ended empirical research: it forces you to listen to what the gathered material has to say. When I was reading the interview transcripts, the emotional responses to algorithms were by far the most interesting aspect of the material. They suggested that our technology relations have become affectively charged in a new way. We feel excited, afraid, and frustrated in algorithmic relations, often without knowing exactly why.

Algorithm-related feelings were not an entirely new theme for me. I had discussed the emotionally charged nature of algorithm talk when I presented initial research findings in public discussions to various audiences, including fellow researchers, technology and communication professionals, policy makers, and other interested gatherings. These talks triggered animated discussions; people could relate to the feelings I was talking about. They too had felt pleasure, fear, distrust, frustration, and irritation in their everyday algorithmic encounters. Yet I had avoided digging deeper into the emotional aspects present in my empirical material. Inadvertently, I had internalized a larger trend that treats technical, political, economic, social, and ethical developments as important and worth exploring, while downplaying emotionally charged responses to algorithms. Despite decades of research on cultural patterning of emotions, it is common to think of emotions as detached from ethical ambivalences and political-economic developments. By using the collected interview material, I could push back on the trend of separating feeling and being from knowing; delve into the emotional-experiential realm; and use personal experiences to highlight the shared understandings of, and reactions to, the everyday manifestations of corporately produced algorithms. This could reveal important similarities between disparate algorithmically mediated practices, such as dating, listening to music, tracing sleep, and reading the news. Furthermore, I could tie these practices to the broader landscape of informational asymmetries.

The Feel of Algorithms explores a cultural shift in contemporary society, one that promotes affectively charged technology relations that most of us will have to acknowledge and account for in one form or another. The study of emotions offers an intervention in discussions that ignore everyday responses to algorithmic systems: how people promote, evaluate, and act in relation to them. Such an intervention, this book argues, is needed to balance the current discussion, which has a tendency to draw conclusions about algorithmic technologies based on celebratory or oppositional responses to imagined future effects. An everyday focus zooms into experiences of pleasure, fear, and irritation, highlighting how the political aims and ethical tensions play out in people's visions, practices, and emotional responses.

Acknowledgments

This book would not have been possible without a generous group of students and scholars who have supported the reported research in its different phases. Most of the interviews discussed in the book were conducted by my research assistant Julia Granroth. She has a natural talent for getting people to talk about what matters to them, and her interviews, conducted in 2017 and 2018, offered a firm terrain for thinking about what goes on in people's lives in relation to algorithms. Other research assistants—Noora Hakkarainen, Ilona Hoikkala and Mea Lakso—benefited from Julia's carefully crafted field reports as we could discuss how to strengthen the empirical material we already had before they conducted additional interviews. Kirsikka Grön and Kirsi Hantula conducted the final interviews for this book as part of the Everyday AI project funded by the Foundation for Economic Education (2020). These interviews alerted us to the theme of algorithm fatigue, raising questions that led to the focus on irritation and frustration in algorithmic relations.

Juuli Hilska, who joined the project in 2020 after I had decided to write this book, helped in the analysis of the interviews and offered support and inspiration when I doubted the rational for writing about "feeling" algorithms. Juuli was convinced that this is exactly the kind of book that people would like to read. Time will tell if she was right, but her many suggestions and our related conversations definitely made this project more fun and the book more readable.

My research group at the University of Helsinki, working in the Kone-funded Algorithmic Culture research project (2019–2022)—Tuukka Lehtiniemi, David Moats, and Sonja Trifuljesko—has kept the debate going on how to think about algorithmic developments without reducing them to top-down political-economic developments. Together, we have been able to identify strengths but also gaps and shortages in earlier research concerning algorithmic systems. I also worked with Maiju Tanninen and Turo-Kimmo Lehtonen on aspects of algorithmic culture that have to do with insurance, which contributed substantially to this book by underlining the historical and cross-sectoral differences of algorithmic systems. Kirsikka Grön, Laura Savolainen, Frank Trentmann, and Domen Bajde have commented on chapters of this book, offered valuable critiques, and enriched my thinking concerning algorithms in the everyday. Conversations with Oskar Korkman and Heli Rantavuo have also been helpful in this regard. Jesse Haapoja read the whole manuscript and suggested insightful corrections. Finally, Marie-Louise Karttunen, my dear friend and the most brilliant language editor that a non-native English speaker can have, whipped the book into shape by making structural suggestions that pulled the different threads of the conversation together. I am grateful to senior editor Michelle Lipinski and her team at the University of California Press for offering the final guidance.

The thinking behind this book has been influenced by collaboration with Dorthe Brogård Kristensen, Mika Pantzar, and Julia

Velkova, who also provided excellent comments and improvements for the manuscript. Natasha Dow Schüll questioned some of my concepts and framings in a thought-provoking manner. Dawn Nafus made me rethink the introduction with her engaged and critical remarks. I am thankful to have had such excellent scholars as companions when trying to capture the current sociotechnical moment. The research network on self-tracking with Martin Berg, Vaike Fors, Deborah Lupton, and Sarah Pink offered valuable ideas on how to expand the scope of self-tracking research to the larger field of sociology and anthropology of algorithmic systems. Our joint work led to another inspiring research network on rehumanizing automated decision-making and a shared book project on everyday automation. Many of the ideas developed in this book and in our network have seamlessly supported each other. I also want to thank Anne Kaun, Helen Kennedy, Stine Lomborg, Kaarina Nikunen, and Susanna Paasonen for their own work and related conversations that have helped to clarify my thoughts. This book is an attempt to strengthen the voices that all these accomplished colleagues have brought to this conversation. I look forward to hearing your views on the outcome.

My family has suffered through the writing stages, regularly asking whether I will be working on the book—again—over the weekend. Now that this project is done, it is finally time to focus on the everyday beyond algorithms. Luckily, there is still a lot of that too.

Introduction

When people recount unpleasant experiences with algorithms, they have a story to share. A fifty-year-old mother and practical nurse, whom I will call Maisa, described how one of her children broke an ankle, an event that she shared in a Facebook update. Later, by mischance, the same thing happened to her second child, and she wondered fretfully in an update how such bad luck could be possible. That same day, an insurance salesperson contacted her and asked whether she would like to obtain additional insurance coverage against accidents, "because an ankle may break." Maisa pondered whether the insurance company had somehow learned about the accidents that she had shared online, thus highlighting the uncertainties connected to algorithmic operations.

The lack of certainty relates to the difficulty of knowing what algorithms and people behind them actually do. Typically, when interviewees describe their responses to algorithms, they are not on firm ground; even professionals with the practical skills to steer algorithmic operations are often perplexed when thinking about their organizational implications. Even if growing numbers of algorithms are open source, some of the most influential ones are treated as proprietary knowledge, veiled for reasons of corporate and state secrecy. Professionals in cybersecurity or digital marketing, actively gathering up-to-date evidence about algorithmic operations, have to work with partial information. Google Search, for instance, is updated regularly, with consequences for the online visibility of companies and organizations around the world, yet organizational representatives argue that concealment prevents abuse via manipulation that might "game" the algorithmic system and jeopardize its functions. And they are, of course, not mistaken; there are many reasons for trying to game and influence algorithmic operations, if they are closely connected to monetary gains (Ziewitz, 2019).

In addition to the lack of certainty, Maisa's story raises the question of the truth-value of algorithm talk. The story told about her children's broken ankles and the subsequent call from the insurance company might not be strictly factual; even if it is, given current regulations, a Finnish insurer cannot use what people write on Facebook for personalized marketing (Tanninen et al., 2021). In interviews, people tell stories, including urban legends, to emphasize something of importance to them. Personal algorithm stories can fail to separate fact and misconception, and they might be based on wishful, erroneous, or fearful views of what is going on. Yet rather than treating algorithmic folklore as evidence of ignorance or misguided reliance on simplified cognitive heuristics, this book suggests a different approach. We will enter the realm of voices and knowledges of vernacular culture (Goldstein, 2015). Instead of concentrating on how people fail to comprehend algorithmic operations, the analysis takes the difficulty of uncovering algorithmic logics as its starting point. The not-knowing, or only partial knowing, explains why personal anecdotes have become such an important source of algorithmic knowledge. We get to know algorithms by feeling their actions and telling stories about them.

Technically incorrect, imprecise, and unsubstantiated comments about algorithms can leave technology experts rolling their eyes. They might insist that we need to define what we are talking about: algorithms are recipes for technical operations, instructions for carrying out tasks and solving problems. Technically, the Google algorithm is not one algorithm at all but countless subalgorithms, each of which carries out a specific task. Factually, algorithmic systems are characterized by a complex and dynamic interplay of multiple algorithms with different aims, assembled by various professionals and engineering teams. Personal algorithm stories, however, are occupied less with technical details than with expressing and translating algorithmic experiences. Nick Seaver (2019a, p. 419) defines algorithmic systems as "dynamic arrangements of people and code," underlining that it is not merely the algorithm, narrowly defined, that has sociocultural effects, but the overall system. Remarkably, as Seaver (2017, 3) points out, many of his interlocutors in highly technical settings could offer technical definitions of algorithms, but they would also talk about various properties of a broader algorithmic system in vague and nontechnical ways. One of the engineers insists that "algorithms are humans too," referring to the human-machine connections that algorithmic systems generate. What people think that algorithms are and what they connect and do matters more in terms of algorithmic culture than precise definitions, because those ideas become part of everyday understandings and personally felt experiences of algorithms. When we do not know the technical details of algorithmic systems, the way we react to algorithms and describe them becomes more crucial in terms of the feel of algorithms than factual or balanced accounts. If we think that algorithms are humans too, we treat them differently than we would if we regarded them as merely parts of machines.

Data Is Power

It is no coincidence that Maisa thinks she might have been observed by the insurance company on social media. Personal algorithm stories resonate with broader shifts in society that have made questions of surveillance newly relevant. Across various domains, in fields from media to health, in political life and the private sphere, the tracking and surveillance of actions and activities is expanding and becoming ever more fine-grained (Pridmore & Lyon, 2011; Zuboff, 2019; Ruckenstein & Schüll, 2017). Jose van Dijck argues (2014, p. 205) that "dataveillance"—referring to modes of surveillance that monitor users through social media and online communication by means of tracking technologies—penetrates "every fiber of the social fabric," going well beyond any intentions of monitoring individuals for specific purposes. Dataveillance is a product of the accumulation of data by the machinery of corporate marketing, including the harvesting of digital traces—likes, shares, downloads, and social networks—that have potential economic value (Zuboff, 2015). The capacity to analyze behavioral and geolocational data with the aid of algorithmic techniques and large volumes of quantitative data suggests "a new economic order that claims human experience as free raw material for hidden commercial practice of extraction, prediction and sales" (Zuboff, 2019).

Everyday algorithmic encounters speak to the intensifying logic of datafication, referring to "the ability to render into data many aspects of the world that have never been quantified before"

(Mayer-Schönberger & Cukier, 2013, p. 29). Datafication is related to digitalization, which promotes the conversion of analog content, including books, films, and photographs, into digital information. As new forms of datafication deal with the same sequences of ones and zeros as digitalization—information that computers can process—they are often discussed in similar terms. Datafication, however, is closely linked to political and economic projects, thereby setting the scene for more general trends and concerns in the current sociotechnical moment. The intensification of processes of datafication suggest that everything about life that can be datafied ultimately will be.

Nick Couldry and Ulises Mejias (2019) frame ongoing developments with the metaphor of "data colonialism," which resonates with how local experiences are being subordinated to global data forces. Data colonialism introduces an extractive mechanism that works externally on a global scale, led by two great powers, the United States and China, but also internally on local populations in different parts of the world. The powerhouses of data colonialism, including Google, Microsoft, Apple, Facebook, and Amazon, aim to capture everyday social acts and translate them into quantifiable data, to be analyzed and used for the generation of profit. Hardware and software manufacturers, developers of digital platforms, data analytics companies, and digital marketers suggest that a growing range of professionals is taking advantage of the datafication of our lives in order to colonize them. Indeed, Couldry and Mejias (2019, p. 5) conclude that data colonialism equals "the capitalization of human life without limit."

Given the informational asymmetries and economic pressures, it is not surprising that algorithms are associated with grim and dystopian predictions of the future. Further critiques of algorithmic mechanisms address how biased algorithms favor privileged groups of people at the expense of others; algorithms discriminate, are not accurate enough, or fail to provide the efficiency they promise. The harms connected to algorithms are also associated with distorted and fragmented forms of self and sociality in families and in peer groups (Turkle, 2011). Natasha Dow Schüll (2018) argues that the intrusive nature of commercial activities can corrode our self-critical capacities and individualize us to the degree that the social becomes dissolved. She describes a vision of "frictionless living" that guides technology designers in their aims "to gratify us before we know our desires." All these concerns are present when people reflect on and evaluate what algorithms do. Algorithmic technologies seek to become intimately involved in the everyday through a novel approach that treats life as minable potential, taking advantage of the monitoring of real-time behavior. Not only are people's lives becoming a source of data, but that data is being used for economic and political purposes in ways that have not been possible before. Digital services, taking advantage of data and algorithms, combine the commercial and noncommercial, the intimate and surveilling tendencies of algorithms, and trigger questions about who is guiding and controlling whom and what needs regulation and protection.

Introducing Friction

Critical political-economic analysis explains shifts in power and profit-making strategies, but it deals only superficially with the question of why tracking technologies are tolerated and even embraced despite their larger political-economy context, privacy threats, and opaque forms of datafied power. This book introduces people like Frank, a growth hacker, whose goal is to make digital marketing more effective. He is inspired by Alexa, Amazon's voice-controlled digital assistant that, ideally, learns what he wants after a few completed purchases and searches preemptively for the cheapest possible product options. What a relief it would be to have everyday necessities like detergent automatically procured! Frank would willingly give up the private information needed in order to outsource tedious everyday tasks to an automated domestic servant and get household goods delivered with little effort. He believes that the more information he provides about himself and his behavior, the more the digital system learns and the better the services and advertisements he receives.

The notion that digital services, boosted by data and algorithms, provide ease and convenience expresses long-standing thinking about the role of technology in society (Tierney, 1993). The historically rooted vision of machines speeding things up and taking over dreary errands that require little or no human skill is a notion commonly shared by professionals when anticipating algorithmic futures. Frank imagines how, by sharing data traces and being as informationally transparent as possible, we can benefit from algorithmic operations. He considers algorithms to be a necessary part of digital life, as they help to navigate vast amounts of information swiftly. Why should we be afraid of algorithms that support us at work and in hobbies, promote sociality by bringing like-minded people together, help us to catch the right bus, predict local weather conditions, and diagnose serious diseases?

If we want to understand the generative nature of algorithmic culture, it is not enough to conclude that Frank is a product of current neoliberal political-economic conditions, co-opted by company promises of data-driven convenience. Instead, we need to explore opinions and values that we do not agree with and reflect the coexistence of anxiety and routinized utility. The ambivalence that accompanies reactions to corporate uses of personal data calls for approaches that do not try to smooth tensions away but can comfortably address the contradictions and balancing acts involved. Personal responses to algorithms engage with this balancing when they hover between positive and negative evaluations of algorithmic developments.

I began formulating the everyday approach to algorithms with the notion of "friction," introduced by Anna Tsing (2005), to engage with how global processes shape the local and vice versa. Friction is also a term used by engineers and designers when they seek to develop perfect human-machine loops. Their aim is to reduce friction and tie people to machines. Frictionless living with computational tools, a man-machine symbiosis, in which the human is unaware of being gently directed by forces of automation, is the ultimate accomplishment (Schüll, 2018). For Tsing, however, friction is not related to a techno-symbiotic dream; rather, it is a societally attuned and resilient notion. Friction makes connections influential and effective, but it also "gets in the way of the smooth operation of global power." In light of friction, globally wired, data-extracting machinery is not exactly the well-oiled apparatus it is often imagined to be. If we believe that human life can be capitalized on without limit, we are giving far too much credit to current data technologies and far too little to the human agencies involved.

Originally coined for the purposes of understanding how global connections sustain claims of universality by becoming locally reconfigured, the notion of friction aids in addressing the tensions and contradictions involved in processes of datafication and related informational asymmetries. We can detect traces of dataveillance in remarkably different places. Yet processes of data extraction are also defined by gaps and breakages that continue to matter (Pink et al., 2018), and it is important not to approach processes of datafication within a predefined, universal framework (Milan and Treré, 2019). Tsing observes that in order to become universally appreciated, concepts and ideas need to travel across differences. Technology-related developments are exemplary in this regard as they mobilize people and organizations in strikingly different societies, from China to Israel, the United States to Russia. People in the wealthier parts of the world anticipate and prepare themselves for impending futures with abstract concepts like big data and artificial intelligence (AI), and when individuals and organizations pick up these concepts, work with them, and affirm them locally, they pave the way for technologized futures. Locating experiences with algorithms within the economic, political, regulatory, and ethical frameworks with which people are most familiar and see as worth pursuing clarifies what excites, troubles, and moves them in algorithmic developments. Frank, for instance, is not only inspired by the convenience of the digital assistant; he is also ready to experiment with the latest technologies. His enthusiasm works as an everyday engine of algorithmic developments.

Viewing datafication and algorithmic technologies through the lens of friction suggests that their powers should not be taken for granted or treated as isolated from mundane experiences and practices. Tsing describes how friction shows us "where the rubber meets the road" (2005, p. 6). The respondents of our study are typically not acting against datafication, nor are they escaping it altogether; they might not even want that. Yet their everyday uses of algorithmic technologies are still not as uncritical and straightforward as the companies or their opponents might suggest, and the friction involved reveals ambivalences and contradictions in algorithmic culture, maintaining a sensitivity to mutable circumstances of life. While a sole focus on the political-economic aspects of datafication can distort the perspective on the everyday, simplify how algorithms are felt and accommodated, or ignore lived experience altogether, incorporating the notion of friction into analysis calls for careful examination of the links between universally appealing goals, processes of power, and locally rooted aims and practices. Thus the friction approach never strays far from the experiential realms in which processes of datafication become personally and societally felt invitations to participate in global developments. The fact that algorithmic awareness leads to more active engagements with digital services, for instance, needs to be taken into account, as it suggests that such involvement strengthens feelings of mastery in relation to technologies (Eslami et al., 2015). Those who trust their digital skills feel that they have agency in digital environments. Unsurprisingly, then, the belief that technologies aid in making everyday lives more convenient resonates most with professionals like Frank who are enthusiastic and skillful in their technology relations.

Finland as an Exemplary Site

The research that led to the study of friction in relation to processes of datafication took place in Finland, where digital technologies feature in future strategies and publicly funded projects that try to anticipate how society needs to be rearranged and citizen skills to be updated in order to thrive in the algorithmic age. Most people in the world know very little about Finland, a parliamentary republic of around 5.5 million people located between Sweden and Russia, whose level of education is high by international standards, which helps to explain the generally good understanding of algorithms. Finland is also the most sparsely populated country in the European Union—one of the drivers of digitalization, as public service delivery can triumph over long distances with the aid of digital services.

As one of the most digitalized societies in the world, Finland actively promotes data-related developments. The national self-image has been techno-oriented at least since in the end of the 1990s, when Nokia's mobile phones were integral to the project of being at the forefront of the global scene. Unsurprisingly, then, public sentiment connected with algorithms leans toward the anticipatory and hopeful rather than the concerned and critical. Despite the optimism, however, algorithms continue to trouble, because they contain and are associated with foreign powers, insecurities, and unknowns. Since Finnish developments and discussions offer ample material for the exploration of the tensions and ambivalences involved in algorithmic culture, this book uses the lens of friction to explain how Finns can simultaneously pursue and find troublesome the deepening of datafication and associated expansion of algorithmic relations.

A governmental goal in Finland has been to pool society-wide resources to foster advances in AI and automated decision-making. The ongoing AI program, for example, consists of initiatives to boost economic growth by revitalizing collaboration between companies and the public sector. Civil society organizations, on the other hand, are concerned about the uses of the data that underpins automated decision-making. Some of the most vocal critics are technology professionals seeking alternatives to exclusive and opaque data gathering and analysis, calling for more regulation, and praising the General Data Protection Regulation (GDPR) enforced in the European Union. A key actor, whose work has influenced many of our interviewees, is an international nongovernmental organization (NGO) called MyData Global, which grew out of a Finnish data activism initiative. Advocates of MyData underline that current business models, with patents, trade secrets, and companies jealously guarding their databases, are blocking healthy digital development, and alternatives are urgently needed (Lehtiniemi & Ruckenstein, 2019).

Surveys that measure public trust repeatedly place Finland among the top countries globally, and it is customary to be confident that governmental agencies are keeping the institutional foundations of society secure through a range of measures that include how they handle forms of data. Longitudinal data sets and statistical analyses have been a self-evident characteristic in the building of the welfare state, in schools and hospitals, in the operations of the tax authorities, and in the criminal justice system. Given the high level of public trust and commitment to openness, Finnish society has a lot to lose with the expansion of digital developments characterized by the use of proprietary algorithms and associated concealment and opacity. With their capacity to track everyday movements and behaviors, algorithmic systems depart from traditional arrangements for handling data about citizens and securing desired societal foundations. Thus data gathering and use cannot be confined to predefined practices in the same ordered way as before, suggesting disruptions and redefinitions of institutional practices. The concentration of data power and the introduction of automated systems can lead to negative social impacts that undermine democratic processes, strengthen inequalities, and deepen poverty (Eubanks, 2018; Dencik & Kaun, 2020). Furthermore, the introduction of algorithmic systems can result in a reduction in public services, as the goal of automation is typically to shrink the amount of human work needed. Driven by pressures to do more with less funds, public sector authorities across the board are employing and planning to deploy automated procedures in their service delivery, even if the absence of humans can leave the most vulnerable in society without sufficient care and attention.

While Finnish developments, particularly in the public sector, follow the general trend of automation—of doing more with fewer human resources—amid these changes one can also sense a genuine attempt to protect the ideals of good governance that define a Nordic welfare society. This tension is important to acknowledge, as it can offer thought-provoking comparisons with developments elsewhere. Public-private partnerships raise fundamental questions about who is guiding whom, as the careful management of such partnerships in the delivery of public services is critical in terms of maintaining the trust citizens feel in state institutions and the welfare society. Since citizens cannot feasibly opt out of the technical infrastructures that underlie public administration, they have few opportunities to participate in the deliberation of values promoted by digital infrastructures (Dencik & Kaun, 2020, p. 3). When decision-making is shaped by algorithms and automated systems, public values require preemptive protection. This means that in order to sustain public trust the Finnish administration needs to preserve and promote the values of good governance proactively. For example, in the care sector companies typically offer algorithmic models to spot correlations between individual risk factors. The implementation of such risk models, however, could fundamentally change the logic of social and health-care systems, a possibility that needs to be carefully considered.

From the perspective of tensions between global developments and local responses, then, Finland is exemplary with its values of social justice, equal opportunity, and a strong educational base for all. In the United States, for instance, people often think of Finland as a socialist country, because it is unimaginable that a mostly free-market economy with a high per capita GDP could be combined with an extensive Nordic welfare state that actively alleviates social inequalities. The positioning of Finland at the forefront of current digital developments, but with limited means to argue against broader political-economic trends, brings out the ambivalence that accompanies attempts to be part of the global scene while realizing that something valuable might be lost locally in the process. This, I argue, makes Finland an excellent site for thinking about the tensions that define algorithmic culture, attempts to resolve them, and alternative ways forward.

Most Finns who shared their experiences in the interviews consider maintaining a successful balance between harmful and beneficial technology developments an important goal. Problematic developments are commonly seen as being initiated elsewhere. Discursively, this kind of positioning—typical of countries and communities for whom protecting their own interests, boundaries, and values is high on the agenda—defines Finland as being threatened by forces of datafication that originate from outside; everyday experiences resonate with the notion of data colonialism, as the country is envisaged as being overtaken by larger, "foreign" powers. The critique continues to be selective, promoting very particular ways of discussing algorithms; the critical perspectives voiced by Finns in the material that led to this book focus predominantly on the services of large data companies, Facebook in particular. In addition to condemning the normalization of excessive data gathering and surveillance, the Facebook critique covers the negative impacts of the platform in terms of democracy and public culture at large. The operations of powerful data companies are challenged, while locally initiated automation operations and AI strategies are left intact. Digital disruption, then, is seen as a good thing if it advances favorable societal developments, but when profit is pursued at the expense of citizens and consumers, technological developments begin to trigger calls for amendment. Finns talk about how they would like to see algorithmic systems governed, regulated, or even banned, proposing that such systems invite us to confront and engage with ethical, political, and societal matters, on both individual and societal levels.

Living the Metrics

This book approaches algorithmic culture through experiences of the ordinary. It tries to capture the joys, concerns, and troubles that are personally felt, because they remind us that people do not reside somewhere outside of the forces of datafication. I started probing issues related to algorithmic culture with a focus on self-tracking techniques used to display personal data about physiologies and everyday behaviors in the quest for self-knowledge. With smartphones and watches, the tracking and measuring of aspects of lived lives have become everyday practices; people use their devices to check on steps taken, hours slept, and distances traveled. Self-tracking research offered the opportunity to attend to how consumer devices and services either become, or fail to become, integral to processes of self-making and how are they linked to broader political and economic processes (Nafus & Sherman, 2014; Nafus, 2016; Lupton, 2016a, 2016b; Ruckenstein & Schüll, 2017). Our team at the time documented how heart rate variability measurements transform physiologies into information and feed this information back to people as scores and charts about their performance and practices, enabling and promoting sensory and informational attachments to daily doings, including walking, sleeping, drinking, and recovering (Ruckenstein & Pantzar, 2015; Pantzar et al., 2017). We demonstrated, for instance, how the visualized heart rate measurements convert sleeping into an arena of observation, thereby triggering new kinds of affective ties between people and their sleep. The monitoring of the quality of sleep through heart-rate variability measurements can deepen care for, but also disciplinary relations with, one's sleeping body. When tracked, sleeping becomes an activity, even a competence, that people can feel that they are good at (Ruckenstein, 2014).

With the experiences gained with self-tracking, I started to observe a broader field of interactions—which I later termed "algorithmic culture"—in which people's visions, ideas, aims, and behaviors become shaped by algorithmic systems. Following this notion, relations can be defined as algorithmic when connections to self, others, and the larger society are seen as being mediated by algorithmic systems. In addition to engagements with numbers and scores, algorithmic relations promote feedback loops that the various metrics support. The notion of the feedback loop can be traced back to the cybernetics of Norbert Wiener (1948) and later applications in the fields of psychology, epidemiology, military strategy, engineering, and economics; with algorithmic systems, they are proliferating. Practices such as following likes on Twitter, buying books on Amazon, listening to music on Spotify, choosing partners from a dating application, or sharing cycling routes in Strava all take advantage of data collected about people and fed back as suggestions and recommendations with the purpose of shaping and modifying routine practices and choices.

In light of algorithmic experiences, feedback loops are worth investigating, because they have become a defining feature of how we interact with digital devices and services. Technologies aim to be responsive; they want to become participants in people's lives by promoting user engagement and the understanding that the everyday consists of arrays of numerically defined practices that can be iteratively examined and acted upon. Scores and charts that use data as their material act back on people, creating feedback loops that figure in intimate experiences with technological systems. Both public and private sector organizations seek to improve products, personalize services, and target information more effectively with the aid of feedback. The expectation is that the feedback information creates connections based on data about previous experiences and then modifies the service experience by means of those connections. Schüll (2016) describes how technology developers and marketers of personal health technology design products to assist and reinforce chosen behaviors. She detects a thermostat-like logic—another kind of feedback loop— that actively aims to regulate users of self-tracking devices via automated prompts, such as taps and buzzes, to make the recommended choices.

By uncovering the designer assumption that human senses alone cannot handle the insecurities of daily lives, we get a view of how algorithmic practices aim for the interpenetration of technological and human forces and agencies. A rich body of research in fields ranging from science and technology studies and philosophy of technology to anthropology and media studies explores the interpenetration of technological and human forces and agencies, examining the constant coevolving, coupling, and mutual retuning of human subjects and their technological companions (Hayles, 2006; Latour, 2005; Lupton, 2016c; Suchman, 2007). As digital technologies merge with everyday lives, we need to remain attentive to the extent and manner in which those lives are algorithmically mediated, modified, and made—the subject of a collaborative project with Dorthe Brogård Kristensen that studied how self-tracing technologies facilitate perceptions and action, and condition experience (Kristensen & Ruckenstein, 2018). By examining how algorithmic technologies energize people's aims, but also how they might be experienced as limiting the everyday, we demonstrated that self-tracking technologies become participants in the transformation of self-experience; some aspects of the self are amplified while others feel reduced and restricted.

Martin Berg (2017, p. 6) observes that designers of self-tracking devices approach users "as vulnerable beings in need of assistance, advice, and actionable guidance," taking advantage of personal data streams with the goal of slicing life into controllable and actionable units. Here, the exploration of algorithmic relations connects to sociological and anthropological studies of quantification, exploring how life becomes structured by means of metrics (Espeland & Stevens, 2008). The study of numbers and measurers and their defining and predictive logics is not new (Strathern, 2000), yet it is an increasingly timely effort. With the spread of algorithmic systems from fields of media and health to finance and education, scores and charts have drifted out of the computational realm to shape the everyday. Credit scoring, hiring practices, allocation of social benefits, social media engagement, health-care diagnostics, and student evaluations can now rely on algorithmic logic that makes recommendations and reaches conclusions without human involvement.

Digital services teach us to cherish quantified information, likes, followers, retweets, and citations and to develop a more affective relationship with numbers and scores. Focusing on Facebook, Carolin Gerlitz and Anne Helmond (2013, p. 1360) argue that within the like economy, "data and numbers have performative and productive capacities." Pedometer users treasure reaching the daily target of ten thousand steps, even if they have no clue about the rationale or justification behind that number. Researchers can initiate their workday by admiring the number of publication citations they have in Google Scholar or peer-to-peer services like ResearchGate. Numbers acquire qualities that promote affective engagements. Yet what is even more crucial in terms of algorithmic culture is that these engagements promote novel practices; we do things in a new way because we feel its effects. We are living the metrics. This is where new patterns of experience begin to emerge, and people think differently about themselves and act accordingly. Algorithmic relations take hold of the everyday bit by bit, so that our recognition of what is happening often trails behind; we notice afterward how our behavior, or some aspect of our lives, has become modified by a specific metric technology. Pedometer users develop new kinds of walking-related activities to increase their step count, but sometimes they also get the numbers up by merely shaking the device. Journalists craft stories according to the metrics used for evaluating their performance. Social media triggers an urge to return to see whether the number of likes is going up; if it is not, one might have to delete the post and try again. Scholars deliberately aim to publish research that attracts citations and create concepts that are particularly citable; influencers who depend on algorithmic visibility for their livelihoods, whether on YouTube or Instagram, post provocative videos to attract attention, or they boost their desirability to advertisers by buying followers.

Avoiding Drama

Numbers, metrics, and feedback loops leave affective marks on everyday experience. We feel anxious when the numbers are not high enough and pleasure when the daily targets are reached. Living the metrics sustains an "affective infrastructure," which "organises modes of presence and participation" (Sharma & Tygstrup 2015, p. 8) in algorithmic culture. The feelings activated by metrics and feedback loops are recognizable, but due to their intimate nature, it is still easy to push them aside and concentrate on the more readily observable aspects of algorithmic culture. Algorithms have quickly moved to center stage in social research, with studies focused on the role of algorithms in financial crises (Karppi & Crawford, 2016), in activism (Velkova & Kaun, 2021), in search engines (Noble, 2018), in mothering (Thornham, 2019), in journalism (Christin, 2020), in digital and physical health (Williamson, 2015), and in rehabilitation (Schwennesen, 2019). Insightful studies have made sense of an abandoned luggage algorithm (Neyland, 2019) and experimented with algorithmically supported walking (Ziewitz, 2017). Amid all this scholarly buzz, however, research that seeks to understand how ordinary people envision, experience, and live with algorithmic systems is still surprisingly limited (Bucher, 2018; Kennedy, 2018; Lupton, 2019; Paasonen, 2021; Pink et al., 2022).

Instead of carefully exploring the interplay between algorithmic technologies and how people contribute to them with their data traces, research efforts have prominently concentrated on the "algorithmic drama," as Malte Ziewitz (2016, p. 4) characterizes research that focuses on the decision-making power of algorithms over our lives and futures. Mundane experiences with algorithms that shape everyday visions and are shaped by them, including questions related to where and when algorithmic systems "feel right" or disturb everyday practices, are ignored when the more spectacular stories of algorithmic powers and wrongdoings take the center stage. My aim is not to deny or undermine the importance of the study of harms and problems related to algorithmic systems. Journalists, activists, and researchers have a critical role in gathering facts and raising awareness about technology-related abuse. I have been part of such investigative efforts myself by collaborating with AlgorithmWatch, a Berlin-based NGO. With a focus on their detrimental aspects, however, important developments that we should also know about might fall under the radar. The oppositional stance is unhelpful if it blocks conversations between various parties and means we no longer attend to other possible scenarios.

I imagine that anybody who studies algorithmic systems from the everyday perspective harbors negative feelings about big tech companies at least some of the time, but it is still important to not to get stuck with adverse effects. In order to learn about the power dynamics of algorithmic encounters, we need empirical evidence that goes beyond the harmful and the undesirable. We need to keep talking to people whose ideas and worldviews unsettle us and do not align with ours. Thus the everyday perspective and friction approach actively reaches beyond the algorithmic drama and calls for formulating engaged critiques and ways forward. Stories of algorithmic encounters reinforce the status quo, but they might also challenge it. Here the study of ordinary experiences is not merely an excavation of reflexivity and individual agency but a project that seeks to understand how to imagine alternative forms of living with data and algorithms (Kennedy, 2018). We need to use our curiosity to discover trajectories that we might not yet envisage; otherwise we will be "absenting ourselves from our futures," as Caroline Bassett and her cowriters put it (2020, p. 2).

Attending to the affective infrastructure suggests that we explore the generative nature of algorithmic relations: how algorithmic culture comes into being in and through the connections people make and maintain. In addition to asking what algorithms are doing to us, we need to ask: What are we doing to algorithms? How are we feeding them with our stories, actions, and engagements? What is crucial is that everyday practices are not merely subject to algorithmic logic; people actively respond to and live with data and algorithms, ranging from actual technical operations to their imagined effects (Bucher, 2018). Taina Bucher's (2016, 2017) work on "algorithmic imagination" provided early steps in this direction by exploring personal algorithm stories that described how specific situations on Twitter drew attention to algorithmic encounters. The novelty in Bucher's approach was that it contested the idea that proprietary algorithms are black boxes that cannot be known (Bucher, 2016). Her methodology relied on the phenomenological approach, suggesting that people tend to encounter the world through "invisibilities"; moods, affects, and values are key in understanding how we experience things and situations around us and learn about them. This implies that people do not need to master the technical specifics of algorithms to know, feel, or imagine what they are or what they might be doing.

The Feel of Algorithms strengthens this line of inquiry by engaging with the stories of citizens, workers, professionals, and civil servants, who are rarely heard in algorithm studies, to demonstrate that in addition to experts, who are developing and using algorithms professionally, and their equally competent critics, those who use algorithmic systems or are targets of them also have the knowledge to evaluate them. As Helen Kennedy (2018, p. 23) insists, "We need to listen to the voices of ordinary people speaking about the conditions that they say would enable them to live better with data." Our interviewees describe in their own words what it means to them that predictive analytics do not revolve around who they are, or that data tells them nothing about what matters to them in life. Their irritation and frustration target the basis of automated decision-making—an amalgam of qualities removed from lived social processes—and also the fact that machine-generated correlations are indifferent to their specific social situations. By attending to how people describe and evaluate algorithmic relations, we begin to see what they think is worth promoting, avoiding, and aspiring to in the technologically mediated everyday. Both fears and future speculations reflect the novelty of algorithmic systems, and new technologies are historically a source of wonder and worry. Algorithmic relations push us to think about what came before, what life was like before algorithms enhanced, manipulated, or distorted our social relations. Fear and enthusiasm are similar responses in the sense that ultimately they might distort the perspective on algorithmic culture, as what happens in the here and now is replaced by what might happen in the future. This poses a challenge for all of us. We need to learn to think about the unruly reality of algorithmic relations beyond the anticipation of good or bad, to pose questions about the sensibilities and habits that come with them, on the go and even before they are fully formed.

Culturally Patterned Emotions

Open-ended discussions about algorithmic encounters gradually grew into a project that suggested that talk about algorithms is a way to carve out different positions in relation to history and an anticipated future. The interviews that we did confirmed the patterned nature of algorithmic talk, with people talking about algorithms in strikingly similar ways. Yet a more remarkable cultural pattern also emerged. When people described their algorithmic encounters, the same feelings, ranging from pleasure through fear to irritation, cropped up repeatedly (Ruckenstein & Granroth, 2020). Neutral and pleasurable emotional reactions are linked to situations in which algorithms operate the way people want, pleasantly surprising with recommendations of music and movies. The emotional landscape shifts into the negative with dataveillance, which appears as disturbance, a feeling that somebody is peering over your shoulder. Finally, emotional reactions narrated as nuisance and irritation are triggered by algorithmic systems operating in too general and mechanical a manner. One might get a suggestion to think about retirement savings even if one is nowhere near ready to retire. Instead of subtly guiding everyday practices, classification schemes become visible as unsophisticated sorting mechanisms that rely on age, gender, and location. The irritation is not directed at the dataveillance per se but rather at algorithmic operations, decisions, and choices that appear too rigid and rule bound.

The affective infrastructure is fed with disappointing experiences of crude segmentation, playing a part in the formation of a combative relationship with algorithmic logics (Lomborg & Kapsch, 2020).

By tracing emotional responses, we can separate realms of the affective infrastructure—the pleasurable, the fearful, and the irritating—and demonstrate the patterned nature of algorithmic culture. The study of cultural patterning of emotions is an established research endeavor, building on decades of comparative research (Lutz & White, 1986; Lutz, 1988). Once viewed as "relatively uniform, uninteresting and inaccessible to the methods of cultural analysis" (Lutz & White, 1986, p. 405), emotions have proven to be powerful research devices for identifying crucial matters at stake in diverse historical, cultural, and political contexts (Hochschild, 1983). In light of this longer trajectory, it is only natural to use emotional responses as a window onto contemporary sociotechnological changes. Taking emotions seriously in the analysis of algorithmic relations emphasizes their epistemological value in knowledge formation (Bericat, 2015; Hochschild, 2002). Emotional reactions are not merely expressions of excitement or fears triggered by a too vivid imagination; rather, they establish a culturally patterned vista, featuring consistent and recurring responses to current processes of datafication. We are dealing with "structures of feeling" that, following Raymond Williams (1961, p. 64), are a living result of algorithmic relations that become manifest in the way people talk about algorithms and respond to their known and unknown effects.

Williams used the notion of structures of feeling for the first time in the 1950s, and over the course of his career he referred to the term in a notoriously slippery and flexible manner. He offered structures of feeling as a framework for thinking about cultural elements that appeared to be at stake when a specific feeling was expressed. Feeling patterns might sustain a dominant ideology, a certain class position, or a patriotic attitude, but they could also oppose hegemonic aims and suggest radically new ways to move forward. Williams was interested in tones, rhythms, moods, and orientations, which "are not exactly signs referring back to an emotional content, but rather expressive building blocks with the help of which a feeling eventually surfaces" (Sharma & Tygstrup, 2015, p. 5). These building blocks mark out the profile of the feeling and its identifiable structure. Fears and mistrust, for instance, resonate with a collectively shared experience of doubt and resistance, despite the fact that fear is often taken to be "private, idiosyncratic, and even isolating" (Williams, 1977, p. 132).

In the realm of algorithmic culture, the point of departure for the analysis of structures of feeling lies in the cultural uniformity of algorithm-related feelings. Although feelings triggered by algorithmic encounters can remain ephemeral and fleeting in the everyday, the way their presence is voiced, privately and publicly, provides important evidence of how algorithmic processes are felt, understood, evaluated, and practiced. Attending to structures of feelings can separate different strands of affectively charged talk and aid in the exploration of friction and how emotional responses to algorithms play out in particular situations. How does experience become articulated in a close and complex interaction between humans and algorithms? How do emotional responses to algorithms maintain structures of feeling? Raising these questions widens the scope of studies of datafication, as they address the everyday production of algorithmic culture through interactions between emotional responses, routine practices, institutions, and power relations. In people's talk, fearful, irritated, and gratified responses to algorithmic developments are aspects of the same dataveillance phenomenon, but isolating the realms of the affective infrastructure can separate dominant and oppositional understandings. This will open novel ways to attend to the emotional alignments and tensions associated with current sociotechnical developments by creating possibilities to study "those moments when new patterns of experience emerge" (Sharma & Tygstrup, 2015, p. 4).

Ethico-political Conversation about the Future

Since structures of feeling do not exist as clean formations in the lived experience, they need to be traced, uncovered, and "purified" for research purposes. The conversations about algorithmic engagements that took place in Helsinki eventually helped us to organize and understand emotional responses that are otherwise easy to naturalize away or take for granted. Asking people to react to what others had said about algorithmic operations produced an ongoing dialogue—a sort of assisted, extended group interview—as one participant reacted to the reflections of another. When dealing with ideas and practices that are on the move, an open-ended and dialogic approach best captures the ambivalent, uncertain, and constantly shifting nature of the phenomenon. It is impossible to document all the details of algorithmic interactions, as technical systems keep changing, adding to the difficulty of offering a stable and up-to-date account. Yet amid the changing particulars we can try to detect cultural patterns that hold together and give internal stability to experiences with algorithms. The structures of feeling are more durable than trends in emerging technologies.

The first interviews suggested the prevalence of fears, pleasures, frustrations, and irritations, thus lending a particular feel to algorithmic culture, but as is often the case in empirical research, realizing the importance of such findings took time. In this case, one could even say that the findings appeared too simple: everybody knows that digital technologies irritate! One of the original goals of our project was to investigate what is algorithmic in culture in light of how people experience and live with algorithmic systems. The methodological stance—that of providing opportunities for observations concerning algorithms with the idea that the most important features of algorithmic culture would gradually emerge—built on a discourse-centered approach to culture that had proven useful for exploring the knowledge claims associated with self-tracking and related practices (Ruckenstein & Pantzar, 2015; Ruckenstein, 2017). Methodological emphasis lay on how talk about a topic, in this case algorithms, encapsulates and valorizes potential uses and meanings for the object being discussed, thereby making it worth engaging with. The study of emotional vocabularies was further supported by the analysis of affective-discursive patterns that arrange social life (Wetherell, 2013). The aim was not, however, to conduct discourse analysis of what people say but to think with the aid of algorithm talk about the broader cultural shift in our technology relations. Structures of feeling are not found in individual emotional responses but in how the feeling structures arrange attitudes, moods, and orientations. The focus is not on how Frank puts his enthusiasm into words but on how his enthusiasm orients him to see what is worth promoting in the co-living with algorithmic systems.

I aimed at forty interviews to create a sturdy foundation for the study of algorithmic encounters. In June 2017 Julia Granroth, my research assistant at the time, and I held semistructured interviews with people of various ages and backgrounds to discuss everyday understandings of algorithms. We interviewed people in their homes, in cafes, or on campus, wherever they felt comfortable, and traced how they develop visions of algorithmic futures and become entangled in data production when they adopt the advice of recommender systems or spend time on social media. In addition to Maisa, who talked about her experiences with an insurance company, the interviewees included students in various fields, as well as a chef, a lifestyle hippie, a postdoctoral researcher, a photographer, a radiographer, a business school lecturer, a nutritionist, a janitor, an internet marketer, and the product marketing manager of a security service. A couple of interviewees were unemployed at the time.

The first interview round, with twenty participants, was followed by two others. In the second interview round seven respondents, who shared skepticism about the current role of algorithms, were invited to shed further light on their avoidance of digital technologies. Sebastian, a law student who is critical of digital developments, talked about how normalization of the excessive data gathering enabled by mobile phones is threatening us as individuals as well as society, emphasizing that anxieties connected to the current scale of information gathering should be taken much more seriously. In the third interview round we studied everyday understandings of algorithms in ten more interviews; for this cohort, frustration might have been a reason to share personal experiences. Since their lack of expertise in questions of algorithmic operations frustrated them, they wanted to think aloud about how they—and people like them—could boost their expertise. They knew that their views might not be correct or flawless, but they wanted to share what troubled them in algorithmic matters as discussion points for others. Here, algorithms began to convert into a "matter of care" (Puig de la Bellacasa, 2017), suggesting that views that are neglected in the current debate need particular attention.

After the three interview rounds, with three interviews short of the target of forty, it started to become obvious that we were not collecting a research sample that would consist of interviews and findings based on them. The research process had moved in a more processual and participatory direction and resembled a much more organic, ethico-political conversation. I started to think in terms of the assisted group interview that brings the different voices together to debate algorithmic encounters. While algorithm talk functions as a realm of conversation and speculation, it includes contemplation and consideration that addresses the question of how to live well in a world where algorithms have become an integral part of the fabric of everyday life. I kept talking with professionals in technology companies about the initial findings of our research. Julia continued with ethnographically oriented interviews that posed questions to digital marketers about the fears and irritations that previous interviewees had connected to advertising. Here the idea was to bring into contact the ideas of those who encounter targeted advertising and those who design and circulate it. Potential research participants were contacted via a Finnish Facebook group called Marketing Collective and by sending interview requests to a business school's mailing list. Locating digital marketers who were interested in a free-flowing conversation about algorithms proved more laborious than finding people who wanted to share critical or suspicious views about algorithms, but using a snowball method, Julia managed to locate nine interlocutors. She also conducted five follow-up interviews with earlier respondents with a background in marketing.

With interview findings and responses to those findings, the algorithm talk started to reveal its organization. In moving back and forth between what people said and how they reacted to algorithms' doings, I paid attention to what was said but also to what was left unsaid. For Finnish professionals, the algorithm talk appeared to be a way to position themselves in the global scene; they want to be in the vanguard of developing new systems, emphasizing that highly educated Finns are well positioned in the global race for technological mastery. Yet they might also feel that processes of datafication have advanced too rapidly, with social norms and regulation lagging behind. For instance, Leo, with experience in a cybersecurity company, pointed out that things are moving ahead with insufficient societal oversight: "We are all part of a societal experiment that we cannot master." When Finns discuss algorithmic systems, they support their observations and claims of how these systems should, and should not, be designed and implemented. This means that ongoing attempts to define and specify commercial, technical, legal, ethical, and societal perspectives become part of the friction that impedes global power.

Algorithm talk positions people in relation to their own histories, including how they locate and see themselves in society and what they think about historically rooted beliefs regarding technologies. Sara Ahmed (2004, p. 26) argues that emotional responses and energies are not merely an individual matter; rather, emotions *do things* that support and shape collective orientations. When describing their experiences, Finns mostly talked about social media and advertising as the location of the algorithmic, whereas many other areas of life where it also features, such as personal finances, predictive policing, everyday mobilities, and the allocation of social benefits, hardly came up in interviews. These absences appeared important, as they obviously reflected local features of algorithmic culture. Devices like credit scoring, for instance, a significant everyday influencer in the United States (DuFault & Schouten, 2020), were not even mentioned. Typically, our Finnish interviewees did not challenge algorithms that aid in deciding whether people can cross national borders, get a loan, find a job, be fairly treated in student examinations, or be sentenced to prison, an indication of their advantaged position in Finnish society. If we had interviewed asylum seekers, for instance, their experiences would have highlighted the exclusionary nature of the welfare society. The oldest of our interviewees were in their fifties, which means that we did not engage with the experiences of the elderly. Based on the interviews, Finnish algorithmic culture builds on the public trust that Finns enjoy and a confidence that they will be protected from the most detrimental real-life effects of algorithmic systems. The histories of racial and ethnic discrimination replicated and possibly accelerated with the aid of algorithmic systems feel distant, and the dystopian developments connected to algorithmic systems are located elsewhere, in China, Russia, and the United States.

Articulations of emotions provide a foundation for what moves and paralyzes, annoys and energizes collectively in terms of data uses and algorithms. While the algorithmic relations that get the most attention are those connected to social media and routinely used digital services, the experiences of Finns are particularly well suited for raising questions about the close interaction between humans and algorithms. In order to bring the discussion closer to actual algorithmic relations, a final round of eleven interviews grounded the conversation in human-algorithm relations. The conversation that we had started in 2017 finally came to an end in 2020, when the pandemic started to mess with us and our algorithmic relations. The closing interviews evaluated the strengths and weaknesses of humans and machines, confirming that the more people engage with recommender systems and digital agents, the more they learn about algorithms and may begin to evaluate and modify their behavior accordingly.

Follow the Algorithm

As confirmed over the course of this book, the algorithm is a fantastically flexible cultural object. It is seen as a resource, an agent, and a buddy; it might refer to a cultural feedback loop, a boss-like authority, a dramatic element, or a symbol of unwanted forces in the digital world and be treated as an influencer, guide, hindrance, and repressor in daily lives (Bishop, 2019; Haapoja et al., 2020; Haapoja et al., 2021; Gillespie, 2016; Lomborg & Kapsch, 2020; Siles et al., 2020). Malte Ziewitz (2016, p. 4) suggests treating the algorithm as a "sensitizing concept," an aid in rethinking deeply rooted assumptions about the politics of automation. Talk about algorithms can reveal what they are thought to be able to accomplish when they promote visions of a better world or bundle together the shortcomings of technology. The algorithm becomes an entry point for addressing an affective infrastructure wherein algorithms are associated with conveniences, ambivalences, tensions, and ways forward. Although we do not think that algorithms have the same level of agency as humans, or even a similar form of agency as humans, talk about them addresses and evaluates whether their actions and functions are seen to have an impact on us and our societies.

Chapter 1 deepens the introduction to structures of feeling in algorithmic culture, demonstrating how affectively charged talk about algorithms unveils cultural understandings and tensions associated with current sociotechnical developments. The realms of the affective infrastructure—the dominant, the oppositional, and the emergent—are outlined for a more detailed study in the forthcoming chapters. A way to study structures of feeling is to search for unifying themes in algorithm talk by tracing what people do, or say they do, in relation to algorithms. The exploration of structures of feeling roots the inquiry in anthropological knowledge formation, which seeks to find similarities and correspondences in activities otherwise thought of as separate. Such similarities are important for the study of algorithmic culture, as they offer evidence of how new patterns of relating to selves, others, and societies might be emerging.

Chapter 2 sets the scene for studying algorithmic culture by way of the dominant structure of feeling, bringing together talk about the neutral and positive feel of algorithms. Identifying the dominant feel of algorithms offers the opportunity to query how feelings maintain and extend algorithmic relations. By doing so, it demonstrates how everyday visions of technology and associated practices establish a foundation for our relations with technology. Experiences with algorithms cover a range of human-technology relations, from the delegation of routine tasks to machines to accepting advice from algorithmic systems on what one should purchase and when to sleep or exercise. This variation is a key to current algorithmic relations, as it underlines the many ways that algorithms "feel right" in the everyday. Some technologies, in some situations, are intentionally kept at a distance, with people insisting that algorithms are merely tools to get things done. This could be the case in a health-care setting, in which the delegation of decision-making to algorithmically controlled systems is seen as a logical thing to do, as they are thought to offer organizational efficiency. Yet when algorithmic systems become an integral part of the everyday, they begin to act back and shape practices and rhythms that define daily lives.

Chapter 3 turns the discussion away from the pleasures of algorithms and introduces the notion of the digital geography of fear to study an oppositional structure of feeling. The goal of the chapter is to relocate experiences of fear from the personal to the collective sphere to demonstrate that the distrust and "mild paranoia" that people talk about is not a glitch in algorithmic culture but its defining feature. Understood in a collective register, fearful reactions become symptomatic of the opacity and uncontrollability of contemporary data relations. Feelings of fear and distrust speak of the impossibility of guarding one's personal space or regulating who peers into one's daily life. An important theme that is highlighted from different perspectives is the lack of information about company operations and algorithmic functions. Company-owned digital spaces are defined by secrecy, thereby promoting the not-knowing that underlies experiences of uncertainty.

Chapter 4 frames an emerging structure of feeling by following the expressed irritation and frustration in algorithmic relations. Articulations of irritation and frustration bring out imperfections, failures, and the controlling propensities of algorithmic systems. When the discussion moves closer to algorithmic operations and the often unrealistic expectations related to them, defined questions about human-machine interactions can be raised. Pinpointing irritation, and staying with it, unlocks a key theme in terms of algorithmic culture: the situational nature of human-machine relations and their everyday consequences. In particular, personally felt irritations offer an evaluative stance that aids in recognizing fundamental differences between human and machinic ways of operating in the world.

Chapter 5 summarizes key findings of algorithmic relations in need of amendment. It revisits those concerned with dominant, oppositional, and emerging structures of feeling, while taking advantage of a conceptual pair: the logic of choice and the logic of care (Mol, 2008). The discussion draws attention to the ethos of broken world thinking (Jackson, 2014), reminding us that our relations with technology are moral and that the rethinking of algorithmic systems calls for ethics of mutual care and responsibility.

Emotionally charged engagements with algorithms challenge us to think about what kind of society we want to live in and who we want to become in the process; presented here, they ask the reader to become part of the conversation. Personal experiences suggest that algorithmic culture is not settled in any way; rather, it calls for engagement. Given the complex nature of algorithmic relations, the conversation does not settle on a single clearly defined, policy-friendly vision for a better future; rather, the various voices suggest different directions by considering the impacts of algorithmic systems on individuals, communities, and societies. It is hoped this will aid in "claiming back alternative futures" (Pink & Salazar, 2017, p. 18). We know that current data-extractive mechanisms, intruding into daily lives and encroaching on notions of privacy and autonomy, are harmful in terms of trust in algorithmic systems, yet in addition to much-needed critique, we need to engage our imaginations and try to find alternate ways forward. Mundane experiences with algorithms that shape and are shaped by feelings and visions, including questions related to where and when algorithmic systems support or disturb everyday practices and aims, are at the heart of this exploration. Algorithmic developments depend on how our actions become data traces and material for algorithmic operations, meaning that the data about us continues to define what algorithmic culture will look like locally and globally, as larger political-economic developments hinge on everyday practices. Nobody will decide on these practices alone or individually, but collectively many different directions are still possible.

1 *Structures of Feeling in Algorithmic Culture*

The shift toward the feel of algorithms is an opportunity to include different kinds of people in the crafting and deliberation of algorithmic futures. It is not only the professional elite that defines algorithmic culture, but also playful, fearful, and inexperienced users testing the limits of current systems or feeling vulnerable and out of place in the midst of digital developments. When people express enthusiasm for and interest in technologically mediated practices, they communicate what they think is at stake in algorithmic relations. Talk about emotional reactions to algorithms does not merely describe how the speaker feels; it valorizes and challenges the object being discussed, thereby rendering it worthy of engagement. Helen Kennedy (2018, 23) argues that in order to acknowledge and enhance our agentic capabilities in relation to data and algorithms, we need "a vocabulary of emotions in researching everyday experiences of datafication." In order to produce such a lexicon, we can trace how emotions triggered by such "everyday experiences" are associated with algorithms by listening to how people make sense of them. In doing so, it soon becomes apparent that when intimate connections are mediated by algorithmic systems, new ways of knowing and feeling come into being that are not just pleasant. Using dating applications, self-tracking devices, and social media, we learn about flaws and weaknesses in others and ourselves that are bound to be distressing. Dating applications suggest one's undesirability when they present no matches, while the pedometer's numbers may inform the user that he is a lazy couch potato. New vocabularies of affective states and emotions are also being created, underlining the need to understand how the emotional connects to collective efforts to live well in datafied times. In Finland, for instance, "heating up" (discussed in chapter 3) refers to a state of agitation, often colored by uncertainty. Addressing personally felt algorithmic powers, it represents local attempts to make sense of processes of datafication, describing particular ways of handling emotional reactions and affects and giving priority to experiential ways of knowing. As feelings and emotional reactions both illustrate the tensions of the algorithmic present and play a role in aspirations for future circumstances, articulations of emotions grow into a resource with which to confront what is going on in algorithmic culture.

From Patterns to Patterning

At first glance, the terms "structure" and "feeling" appear incompatible, as the former implies stability and the latter elusiveness. Bringing the two together, however, is precisely what makes the structure of feeling a useful heuristic device for illuminating the ambivalences and contradictions in algorithmic relations. The analysis of structures of feeling has mainly been deployed in studies of literary and filmic culture but, as Ben Highmore (2016) suggests in an illuminating essay, the approach should be broadened to feelings embedded in habitual, everyday life and shifting cultural phenomena. For Highmore (2016, pp. 145–146), feelings are "agents of history" and "form-giving social forces" that come into being as emotions, tones, rhythms, moods, manners, attitudes, and orientations; we are dealing with patterned emotional responses, detected by means of algorithm talk, rather than the bodily sense of the affective. While articulations of emotions will not provide a direct access to how people operate "as carriers of structures of feeling" (Highmore, 2016, p. 152), talk about them offers an entry point to the analysis of the feel of algorithms, while repeated emotional responses—joys, fears, and frustrations—provide evidence of identifiable patterns in algorithmic culture. For instance, for digital marketers, tackling advertising with algorithmic techniques is much more satisfying than thinking about the risks and harmful societal implications that algorithmic systems might present. Their feel of algorithms leans toward the pleasurable. As with a national ideology that is embodied and engenders patriotic pride, feelings of enthusiasm about the newest automation techniques call into being practices of experimentation and testing. The positive feel promotes the experimental realm in algorithmic culture, which is crucial for its reproduction.

Highmore connects Raymond Williams's attempts to capture moods and orientations via structures of feeling to the earlier efforts of anthropologists Ruth Benedict and Gregory Bateson to analyze how cultural forms and materials convey feelings. In the 1930s, Benedict (2019) promoted the comparative analysis of cultural patterns, suggesting that each culture exhibits a particular ethos that strengthens certain behavioral forms and personality traits at the expense of others. She compared rituals, beliefs, and personal preferences among people of diverse cultures to show that a culture has "a personality" that is encouraged in individuals.

The "culture and personality" school has been fiercely criticized for its essentialist take on culture, but the core idea that pattern recognition forms the basis of cultural analysis is still relevant. Pattern analysis can reveal correspondences in activities otherwise seen as separate; in the realm of algorithmic culture, this includes similarities between algorithmically mediated practices, such as dating, sleeping, listening to music, and reacting to posts online. For the study of algorithmic culture, these similarities are crucial because they offer clues to how novel patterns of experience might be forming.

Bateson (1936) elaborated further on the term ethos, suggesting that it involved shared sentiments and thereby signaled culturally appropriate behaviors. His pattern analysis focused on how disparate practices promoted attitudes, feelings, moods, and manners with an identifiable structure. While Benedict and Bateson studied living cultural forms, Williams traced the longue durée of industrialized change; yet as Highmore proposes, important similarities define their approaches. This suggests that the broadening of Williams's intellectual roots and efforts, with its anthropological linkages, is useful when addressing how structures of feeling materialize in everyday phenomena: from rituals, family, and religion to AI and algorithms. Within this more comprehensive approach, structures of feeling can be seen as an integral part of the ongoing organization of technologically mediated daily life. Trying to capture feelings is the beginning of a journey that queries disparate social forces in the course of trying to unpack the patterning of algorithmic culture, meanwhile moving the emphasis from static patterns to more dynamic processes of cultural patterning.

In this context, the main incentive for focusing on "the feel of algorithms" lies in its openness. The local features of algorithmic culture mean that the feel is not universal but shaped by how people position and see themselves and technologies in society; therefore, before outlining the dominant, oppositional, and emergent forces at play in algorithm talk, I discuss features of algorithmic culture that explain the emotional responses discussed by our Finnish interviewees. The feel of algorithms is inescapably tied to the forms of surveillance and behavioral modification promoted by algorithmic systems. A further source of emotional pressures and anxieties is found in the ways that algorithms connect to administrative and managerial aims, including the commercial co-opting of selves and socialities. Yet what is important in terms of everyday algorithmic encounters is how the broader context becomes known and lived through personal experiences and reflections about them. Here we need to pay attention to algorithmic folklore and the silent pedagogy of targeted advertising.

Infrastructural Interdependencies

Algorithmic systems behave similarly to classificatory schemes that have been used for administrative and commercial purposes for decades, if not centuries: simplifying and standardizing aspects of the everyday with the goal of making consumers and citizens legible (Scott, 1998). Algorithmic technologies promote a formal and impersonal orientation to the world by introducing design-based control and automated rules, thereby appearing as "the latest instantiation of the modern tension between ad hoc human sociality and procedural systemization" (Gillespie, 2016, 27). Algorithmic systems do not treat people as entities with distinct histories and a unique cultural positioning but as data (Cheney-Lippold, 2017). A middle-class female with culture-related hobbies or a man in his forties with an inclination for wine tasting is broken down into machine-readable characteristics: age, gender, Verdi, Tuscany, holiday, wine. Thus, rather than a bounded individual, the algorithmic system is dealing with a "dividual" (Deleuze, 1992) composed of computational features and relationalities. The goal of the commercial system is to detect and combine such features and relationalities to suggest, preemptively, what the dividual might be up to next.

By promoting predictive and prescriptive outcomes in terms of consumer behavior, the market-consumer relationship becomes defined by "an intimacy of surveillance" (Ruckenstein & Granroth, 2020), which introduces new kinds of tensions and ambivalences to the everyday. Whereas the actions of users online remain open and visible to the instrumental control of surveillers, service developers, and marketers, the practices of data gathering and tracking tend to fade into the background for the users themselves. Here the troubling aspects of algorithms are linked to novel interdependencies that digital infrastructures support, which might, however, not be visible to those surveilled.

The societal power exercised by data empires, particularly Google, has been termed "infrastructural imperialism" (Vaidhyanathan, 2012), another formulation of "the external appropriation of data on terms that are partly or wholly beyond the control of the person to whom the data relates" (Couldry & Meijas, 2019, p. 5). Digital services become infrastructural when they provide societal "undergirding," as Brian Larkin (2013, p. 323) puts it. An infrastructure, working as planned, organizes and enables everyday doings to the extent of occupying a taken-for-granted position. Tom, who works as a part-time event planner and entrepreneur, recognizes infrastructural dependences when he discusses his reliance on digital technologies. If there were a technological recession—taking us back to the pre-internet days of the 1980s—his algorithmic skills would simply become obsolete. Marketing without digital tools would require a big shift in know-how. This is a frightening thought to Tom, because with the loss of digital infrastructure, he would lose his capacity to act. Tom concludes that he hopes that the digital will never disappear, as the loss of capabilities would be devastating.

The taken-for-granted role of digital infrastructures means that the feel of algorithms can be "neutral," a term that our interviewees often used to describe algorithms when asked to weigh the pros and cons of digital technologies. In light of understandings of their "neutrality," the work that algorithms do remains unacknowledged as people find information, share experiences and aspirations, and seek an escape from boredom in the entertainment recommended to them (Paasonen, 2021). A manifest feature of digital infrastructures is that they become rooted and normalized through routine use; infrastructures disappear from sight as they mesh with our daily actions. Algorithmic systems orient us socially and affectively, positioning us in relation to others and the world, accentuating what is of interest and value. Recommender systems silently contribute to how we read, what we buy, and the music we play, while in social media informational signaling, including likes, follows, and hashtags, converts sociality into consumable material, meanwhile assisting the navigation of fast-paced information dissemination.

As the digital infrastructure sinks into the background, we might no longer be aware of the difference between active and passive use, or even between use and nonuse (Karppi, 2018; Paasonen, 2018a). Uses of services become so habitual that it is hard to say where they begin and end. What is crucial, then, is that digital infrastructures—and their *neutral* feel—keep current informational asymmetries and power imbalances in place. Susanna Paasonen (2018a) defines mobile connectivity as "a new infrastructure of intimacy" that has important similarities with traditional infrastructural necessities like electricity and water; people in Finland no longer live comfortably without digital connectivity. "Being connected means that one is alive," as one of the interviewed students put it. Data and algorithms are as much a part of the everyday as heating in the home. They define a comfortable life, rather than present an exogenous threat to our culture.

Tensions with Commercial Co-optation

Companies that employ algorithmic techniques to build consumer relationships seek to become participants in the everyday by tracing past and predicting future behavior. As Marion Fourcade and Kieran Healy describe (2017, p, 23), "Increasingly, the market sees you from within, measuring your body and emotional states, and watching as you move around your house, the office, or the mall." Opportunities to profit from ordinary people's lives with the aid of algorithms are plentiful; data on sociability, friendships, fertility, rivalry, sleeping, and breathing is all grist for the mill. Digital developments aid in further blurring the boundaries between corporate interests and daily lives through forms of social networking, sharing, and knowledge formation that promote interactions, collaborations, and divisions of labor between commercial agents and consumers (Bruns, 2008).

The feelings related to algorithms replicate tensions and anxieties present in consumer culture, where moral concerns with the commodification of everyday practices have a much longer history than recent algorithmic systems (Brembeck, 2008; Cook, 2004; Miller, 1998; Zelizer, 1985). Consumer culture is defined and redefined in the ongoing tension between what can and cannot be commodified, highlighting not only clever ways to commodify but also the importance of not painting a totalizing landscape in which commercial agents effortlessly realize specific practices and aims. Digital influencers, who produce content using social media channels to influence others' behavior both online and offline, have generated moral controversy, as their style of marketing expands the realm of advertising to an unprecedented degree (Abidin, 2016). Influencers operate as advertising machines with human faces, personifying "biopolitical marketing" (Zwick & Bradshaw 2016), which blurs marketing with daily life. In the work of digital influencers, the inventive, explorative, and personalized aspects of marketing are particularly valued; the ideal is for the line where marketing begins and ends to become undetectable. Influencers promote commercial messages that are "deeply inserted into, and increasingly indistinguishable from, the fabric of everyday life" (Zwick & Bradshaw 2016, p. 93), responding to calls for marketing to "make life," to better anticipate and fulfill consumer needs and desires. When an influencer talks about life, and commercial products as part of that life, messages should feel authentic and organic as opposed to orchestrated (Abidin, 2016).

Digital services that commodify everyday aspirations take advantage of deeply rooted social aims, including desires for recognition and reputation (Arvidsson, 2005; Turkle, 2011); this is why they feel so intimate. As Sherry Turkle puts it (2011, p. 281), "We warm to machines when they seem to show interest in us, when their affordances speak to our vulnerabilities." Emotional responses to algorithms, particularly frustration and irritation, react to the parasitic and fluid nature of commercial co-optation. Nick Seaver (2019b) traces the influence of persuasive technology on the recommendation system industry and describes how developers speak of getting people "hooked"; users are configured as prey to be trapped by persuasive design. The epistemic resources used by developers in the fields of advertising, marketing, insurance, and digital health often originate in behavioral psychology and behavioral economics. When people talk about algorithms, they can refer to gamification that seeks to influence them by means of game-playing features, including goals and objectives, point scoring, and competition with others. Nudging, on the other hand, is a way to guide users by relying on a "choice architecture" within which they can be persuaded (Seaver, 2019b; Schüll, 2016). Schüll (2016, 323) defines the nudge as "a curious mechanism, for it both presupposes and pushes against freedom; it assumes a choosing subject, but one who is constitutionally ill equipped to make rational, healthy choices."

Forms of persuasion built into digital devices are felt when technologies follow people in their daily doings, aiming to shape and transform them. While the notion of targeted advertising has been around at least since George Gallup's introduction of market research in the 1930s, what is new in terms of everyday algorithmic engagements is that people feel that they must continuously evaluate and counteract technologies in order to maintain their sense of self-determination. Talk about algorithms focuses on the commercial mechanisms that bend digital technologies and us all in the process. "Markets have learned to 'see' in a new way, and are teaching us to see ourselves in that way, too," as Fourcade and Healy (2017, 10) argue. In seeing ourselves like the market, scoring mechanisms and metrics that stimulate affective engagement are of key importance (Stark, 2018). Constant numerical feedback tells how we are doing in comparison to others. Self-worth becomes tied to mechanisms of valuation offered by data analytics: numbers, scores, and rankings. Higher scores are treated as evidence of success in everyday doings, ranging from sleeping and breathing to dating and telling funny stories.

What emerges as pleasurable in personal reflections are rewarding and frictionless algorithmic relations, when technologies are felt as supporting and even caring. Yet even when algorithmic guidance might be seen as convenient and pleasing, it still raises questions about the deeper penetration of technically mediated persuasive forces into everyday aims. Due to their intimate nature, new persuasive practices and techniques amplify the concern that we can no longer protect our thoughts and behaviors from commercial influences. Henna, a student of theoretical philosophy, poses a question that comes up frequently when people assess the intimate powers of algorithms: Who is guiding whom, and based on what criteria? The notion of autonomy converts into an everyday sensor, as it guides the evaluation of what is offered by algorithmic processes and whether they support personally and publicly shared values (Tanninen et al., 2022a). Autonomous agency is assessed and felt in relation to limiting and enabling features of algorithmic systems (Savolainen & Ruckenstein, 2022). Passive consumption of digital services—the mindless following of algorithmic suggestions, for instance—can still feel like active choice; consequently, algorithms are evaluated favorably when they are seen to align with personal interests. From this perspective, autonomy appears as a situational achievement in human-algorithm relations, one constituted by reflective, adjustive, and protective behaviors in relation to algorithms and their imagined effects.

Advertisements with a Wider Resonance

Talk about algorithms tends to stay close to everyday practices and concerns, and targeted advertising triggers critical and annoyed observations related to algorithmic culture. A concrete example is the way algorithms fail to "see" their targets. Sofia, who works as a sales and marketing coordinator, is presented with an ad on her Facebook page about applying for a summer job at McDonald's. She feels undermined, algorithmically bundled into a segment of young people in which, having had a regular job for more than two years, she does not belong. Iida, trained as a nature guide, is equally annoyed. The ads she is shown reflect normative values that the algorithms merely replicate; she is "known" to be of child-rearing age and consequently bombarded with pregnancy-related information. Stine Lomborg and Patrick Kapsch (2020, 754) recount the frustration of a Danish man in his thirties living in a same-sex marriage. He refers to what he defines as "homo-spam" that treats his sexuality as "a stand-alone classifier" for his identity and says that he opposes the oversimplified and stereotyped boxing by means of algorithmic systems. Crudely targeted ads promoting and accelerating stereotypical classifications generate irritation and frustration, but they can also feel like an assault on one's identity, not to mention the dangers they pose when they reveal qualities like homosexuality in countries where people are persecuted based on their sexual orientation.

Reactions to advertisements offer glimpses of how algorithmic culture holds together contradictory forces like intimacy and instrumentality and care and surveillance. Observations concerning ads are not solely about advertising; rather, they spark a much broader conversation that concerns datafication, surveillance, autonomy, market aims, identity pursuits, gender stereotypes, and self-understandings. In terms of how algorithms feel, paying attention to feelings that targeted advertising triggers strengthens the investigation of the more ambivalent aspects of processes of datafication, crucial for thinking about how we live—and would like to live—with algorithmic systems. Since the sorting and scoring mechanisms of algorithmic systems become personally "known" and "felt" through digital marketing, targeted advertising is a prime location for observing how data traces left by everyday behavior are analyzed and fed back in the form of advertising.

What the personal stories about exceptionally well or poorly targeted advertising have in common is that both render algorithmic operations suddenly visible—apparently in possession of private and intimate details—and consequently catch people off guard (Bucher, 2017, 2018). For the most part, however, advertising is an everyday occurrence that goes unnoticed. Ads are necessary clutter that intersect with news sites and social media posts. When asked about the last ad they remember seeing on Facebook, people have to think carefully about what has caught their attention. Yet while most advertising is filtered out, certain ads are noticed and prompt strong reactions, revealing both algorithmic effects and their impact on routine experience. When people feel that their weaknesses have been exposed by the algorithm's "insider" knowledge, it can feel like abuse. Ads feel particularly offensive when they are directed at personal vulnerabilities: a woman suffering with excess weight is presented with dieting programs, a man with a receding hairline is offered hair implants, a young man with an empty bank account gets a lucrative loan offer, and a woman who suffers from infertility is repeatedly reminded of her situation by pregnancy test promotions.

As we compared the annoyed observations triggered by advertising, a pattern emerged of ads being seen to fail to align with personal and societal aspirations and aims. Take Veera, for instance: objecting to fur farming, she reports all Facebook ads that contain fur products as inappropriate. Yet annoyingly, she keeps receiving fur promotions. The algorithmic system appears remarkably clumsy, unable to learn her political convictions despite repeated notification. Generally, however, scholars who study the impacts of algorithmic systems, automated decision-making in particular, tend to focus on facets that are more substantial in terms of real-life effects than advertising. Cathy O'Neil, a data analyst, demonstrates in *Weapons of Math Destruction* (2016) how algorithmic techniques exclude and punish underprivileged individuals and communities by identifying them as potential police targets, restricting their access to financial and health services and reinforcing the racial biases of the penal system. Safiya Noble's *Algorithms of Oppression* (2018) demonstrates that Google's search engine promotes discrimination and racism. Her findings about the hypersexualization of women of color generally, and Black girls specifically, are revealed in search results as well as in the paid AdWords results. In *Automating Inequalities*, Virginia Eubanks (2018) provides an analysis of how automated tools reinforce inequalities and intensify the marginalization of the poor, who face automated scoring mechanisms when they try to access public services. The social costs of receiving poorly targeted advertising are low compared to not being paid social benefits or not being recognized as vulnerable enough to need shelter. In light of such comparison, findings concerning advertising failures can appear banal and frivolous. Yet although these two realms—the allocation of social benefits and targeted advertising—appear disconnected, they draw on shared features of algorithmic culture (Mähler & Vonderau, 2017). Targeted advertising can reveal how mechanisms of automation favor morally contested and societally damaging ways of circulating information. For example, people buying advertising space on Facebook could define their target groups in terms as diverse and controversial as "Jew hater," "brown skinned," "Nazi party," and "recently divorced" (Mähler & Vonderau, 2017). Automated processes of ad buying, mostly lacking human oversight, enable ethnic profiling, terrorist recruitment, discrimination, conspiracy, and the spread of misinformation, underlining the importance of discussing advertising alongside other algorithmic systems. In terms of cultural and societal processes, digital advertising can promote a harmful and dehumanizing logic by treating human features computationally and not caring about the consequences of such treatment.

Algorithmic Folklore

In addition to highlighting the role of automation in directing and allocating information, exploration of how ads and their algorithmic mechanisms feel gave rise to another important finding. While companies use advertising to influence people and push them to make purchase decisions, ads have gained another function altogether: they have become a site of algorithmic pedagogy, promoting a bottom-up form that is neither guided by the advertising companies nor supported by accurate information about algorithmic techniques. The pedagogical power lies in the fact that, by way of advertisements, people witness firsthand how companies use algorithmic functions. With targeted ads, one can trace how data-gathering techniques and data analytics used by companies operate and play out in practice. People acquire knowledge about algorithms as they witness how their characteristics, behaviors, and locations are converted into ads that appear in online sites after online searches, private messages, or even face-to-face conversations. For instance, the similarity-seeking quality of algorithms is easy to detect: if you watch Polish films, you get recommendations for Polish films. When searching online for sofas, you get ads for sofas and occasionally cushions too; you learn that the algorithm connects sofas to cushions.

People get to know how algorithms work by observing what is being done with the aid of data traces they knowingly generate or unknowingly leave behind. In the process, they learn how algorithms prioritize types of content and emphasize certain behavioral features at the expense of others. Anne, a high school teacher, describes her observations of the mechanisms of the attention economy, particularly the privileging of "fresh" information (Gillespie, 2013). Without reactions and feedback, social media posts tend to lose visibility quickly. In order to lengthen the lifespan of her friends' posts—particularly those deserving prominence—she reacts to a post first, then adds a comment, and later returns to read it again. Anne spends time with content that she thinks is worthy of consideration, even if she is uncertain whether this behavior has actual impact. Yet it feels good to express solidarity and support friends' causes.

The pedagogy of digital advertising stresses the role of feelings and reflection—it might be impossible to differentiate between these two—as the foundation of algorithmic knowledge. Observing feedback loops and how traces of behavior convert into advertising teaches the basics of algorithmic logics silently and efficiently, and it might be impossible to articulate how the knowledge has been received. These observations typically take place when we spend time alone with our devices. We are confronted with "digital solipsism", which Mark Andrejevic (2019, 14) describes as a consequence of automation processes. Expectations of constant connectivity promote programmed interactions with digital devices and assistants, moving the emphasis on sociality away from actual human communication. Rather than talking to other people, we reflect on what machines are doing to us and what we are doing to them. People make parallel observations about algorithms, yet they might have no shared arena in which to question and deliberate on those experiences. This gives personal anecdotes particular cultural weight; stories about algorithmic encounters bring into visibility aspects of algorithmic relations that trouble us but might not be shared with anyone.

A typical anecdote could go like this: two women, Paula and Mea, go for a walk on a chilly winter day in Helsinki. They talk about windproof jackets, and Paula mentions a Swedish jacket brand that she wears. The following day, Mea sees an ad on her Facebook home page promoting windproof coats of exactly that same Swedish brand. She knows little about Facebook's data extraction practices, but based on the chain of events, Mea reaches the conclusion that the company had access to their private conversation. The advertisement is only conceivable, Mea tells Paula, because their phones are listening to them. As discussed in chapter 3, it is not plausible that the phones of the two women in Helsinki are actually being monitored. Facebook uses sophisticated demographic and location data to serve up ads; it does not need to eavesdrop on people to get the information it wants. Yet personal algorithm stories can treat eavesdropping as a fact. In that sense, the pedagogy of digital advertising is poor pedagogy, leading people to infer causality where there is no evidence of it. The conviction that phones listen to one's conversations becomes incorporated into algorithmic folklore, informing personal algorithm literacies. These literacies might be based on erroneous or misguided perceptions, but they are algorithm literacies all the same; people read the digital environment and use their observations and gut reactions as pedagogical guides.

Algorithmic folklore is an attempt to control an environment that is in many ways uncontrollable. It marks moments and events when people, lacking adequate information, feel and form opinions on how the algorithmic system works. Not only are proprietary algorithms veiled in company confidentiality, but the functions of algorithms keep changing, making their updates challenging to follow. And even if the technical specifications were open sourced and accessible, it would require specialized skills to understand how algorithmic systems are built and operate (Burrell, 2016). As users of digital services cannot accurately identify the technical details of algorithmic systems, everyday mastery of algorithmic operations becomes crucial. Sophie Bishop (2019), exploring algorithmic gossip among beauty vloggers who depend on algorithmic visibility on YouTube for their livelihood, details how they gather and share knowledge about algorithmic processes by comparing experiences with the aid of media stories and personal anecdotes. Without access to accurate information, gossip about algorithms typically remains unverified, yet it still informs the ways in which vloggers manage algorithmic visibility. As informal chatter solidifies beliefs about the behavior and power of algorithms, it converts those beliefs into knowledge that is acted on, strengthening algorithmic relations and specific features of algorithmic culture.

Ideologies as Sensed Realities

Highmore (2016, 156) asks whether the approach to structures of feeling that he is advocating is merely "a provocation that petitions for fuller descriptions of our collective lives." He speculates that Williams hung onto the phrase "structures of feeling" despite its being constantly critiqued by his peers, because it situated him firmly in the realm of the empirical. Williams was interested in how words and values become registered in feelings to develop an understanding of the reproduction and change of social and cultural forms. Relatedly, "the feel of algorithms" can be used as a context to query distinct social forces associated with technical developments. Structures of feeling are rooted in both cultural and ideological continuities and instabilities; they contain reproductive and resistant aims, tensions, and contradictions that serve as potential points of defiance and confrontation. In light of algorithmic culture, with its unknowns and uncertainties, identifying separate feelings allows the exploration of algorithmic relations that are taken for granted, neglected, and not sufficiently explored.

Feeling good about human-algorithm relations suggests that algorithmic attachments become influential and effective when they facilitate the operations of the globally wired, data-extracting machinery. The convenience associated with algorithms naturalizes the dominance of technologized futures over others, while it solidifies the position of technology elites as vanguard. The personal and societal convenience that digital technologies offer typically appears as an apolitical and ideologically neutral value orientation (Tierney, 1993), yet if feelings are indeed agents of history and operate as form-giving social forces, the ease and convenience that we associate with algorithmic technologies is far from ideologically neutral. Here we begin to see how the feel of algorithms sustains a hegemonic stance that treats algorithms as neutral and benevolent social forces, giving them primacy in thoughts about how practices should be organized and lives lived. Structures of feeling, then, are ideologies that are experienced as sensed realities. They align the personally felt with economic aims and political conditions, thereby maintaining modes of interaction, shared experiences, and emotional responses.

The primacy of the dominant structure of feeling means that it needs to be discussed first before turning to other structures, as it lays the foundations for the inquiry. The dominant structure requires continuous reproduction of optimistic accounts of datafication to maintain its ascendancy, yet retaining a positive response to algorithms entails that negative and harmful aspects of digital technologies are overlooked, downplayed, and presented as solvable—typically with the aid of technologies. The infrastructural imperialism that characterizes the spread of societal influence by data companies is seen as unfortunate, but perhaps there is a way to work with them; perhaps association and collaboration with these companies can help society to prosper as they prosper. In personal reflections the dominant structure of feeling reveals its peak moments in the enthusiasm that supports algorithmic systems. Henrik, a life coach, praises the potential of algorithmic operations to give purpose and direction to human efforts. At its most intense, the excitement features religious overtones, which are, Sally Wyatt (2004) argues, never too far from revolutionary talk about technologies. The belief in the algorithmic outlook offers anticipatory guidance, something that enables forward looking. The feel of algorithms upholds the future promise of convenience and pleasure, and the resulting convenience is a process and an aim rather than a firm end result.

However, cultural aims, even dominant ones, are never completely stable or finished. The neutral and pleasurable feel of algorithms is challenged by an oppositional stance, which is related to the identified harms connected with technology developments, including dataveillance, limited transparency, and commercial co-option. The dominant and oppositional forces are simultaneously present when people weigh the pros and cons and make sense of algorithmic culture. Personal responses to algorithms engage with this balancing when they hover between positive and negative evaluations of algorithmic developments. The emotional landscape is not a permanent state of affairs, but one that is constantly evaluated and updated. What separates dominant and oppositional forces is that whereas the oppositional stance requires documentation, elaboration, and justification, the dominant stance on algorithms is often taken for granted to the degree that it needs no explanation: it is just the way things are, or how they should be. "There is no ideology in progress," as one of the technology professionals put it. In contrast, the oppositional stance, which may call for investigations to demonstrate that algorithms discriminate or are biased against women or minorities, is treated as a political project. The oppositional feel of algorithms is driven by appeals for social justice, by concerns about the disadvantages that technologies have for people's lives and public culture at large.

In addition to the dominant and oppositional structures of feeling, the discussions in Helsinki suggested that articulations of irritation and frustration should be examined more closely and the combination be studied as a structure of feeling. Brita Ytre-Arne and Hallvard Moe (2020) argue in their study of folk theories of algorithms that irritation could inspire future political action against processes of datafication, contextualizing their claim with the notion of digital resignation, suggested by Nora Draper and Joseph Turow (2019) to explain inaction and passivity in the digital environment. When faced with harmful practices, such as privacy violations, resignation is a rational response to the fact that people have insufficient information to foresee the consequences of their actions, and they no longer see opting out as a choice. Ytre-Arne and Moe see traces of resignation in their empirical material, particularly when their Norwegian respondents characterize algorithms as confining, reductive, and exploitative. Yet they argue that the critique of algorithms also conveys emotional engagement—irritation and frustration—that is far from resigned. People might not openly question the existence of algorithmic systems, but they are not resigned to accepting all of their features.

Irritation and frustration, I argue, come together as an emerging structure of feeling that addresses the friction that accompanies processes of datafication. For Williams, the characterization of "emerging" is inspired by the idea that new structures of feeling are constantly developing beyond recognized social and political beliefs. Whereas the dominant and oppositional stances can be found in all kinds of texts, narratives, and projections that deal with digital technologies—technologies are typically either an opportunity or a threat—irritation opens a less-studied perspective. Thus the pleasures and fears associated with algorithms often lead us to discursive either-or positions that feature algorithmic relations as empowering or exploitative. With the emerging structure of feeling, however, we can focus on what might be evolving beyond positive and negative evaluations of how algorithmic practices challenge and rearrange societies. Irritation can be defined as a more persistent "low intensity negative affect" (Ngai, 2005, p. 174), residing in the midst of dominant and oppositional form-giving social forces. Acknowledging irritation moves the focus of the inquiry closer to actual algorithmic operations, suggesting that following the irritation that accompanies our relations with technology unveils the annoying nature of human-machine connections. Idioms are telling in this regard, as irritation can be off putting and "rub us the wrong way" (Ngai, 2005, p. 175). In the following I briefly introduce the three structures of feeling that aid in organizing the affective infrastructure of algorithmic culture before turning in the next chapter to firsthand accounts of emotions that bring their essence to life. The goal is to demonstrate that the analysis of emotional responses opens novel modes of thinking about algorithmic relations, as it can address them in ways that might not become visible without such excavation.

The Dominant Feel of Algorithms

If algorithmic systems work smoothly in the everyday and align with personal and societal aims, they raise few concerns. People might notice what algorithms do, and how digital services behave, but the routine convenience that technologies offer is typically not discussed. The demarcations between data, algorithms, digital services, and the everyday tend to become blurred; in daily experience, they are all part of the same flow of events. In algorithm talk, people do not necessarily mention algorithms at all but focus on the wider digital landscape.

Digital services expand social relations geographically outward, but also topically, when they support the sharing of interests.

The role of digital technologies in enhancing selves and socialities has been extensively explored (Baym, 2015; Rettberg, 2014) and, resonating with earlier findings, Frank describes how digital services sustain personal and social aspirations. He recounts how his WhatsApp relatives' group connects him to rural uncles, aunts, and cousins from whom he learns about fishing, birdhouses, and other subjects outside the experience of his friends in Helsinki. Alongside facilitating novel social ties, the pleasures provided by algorithms are also associated with the deepening of existing social relations. Tom, the event planner, approaches algorithms from the perspective of social networking. In the past he would simply have lost contact with people, but now he can hold onto a friendship, no matter how tenuous it might be, by sending an annual birthday message. By redefining friendship, Facebook creates a reserve of potential "friends" that can be activated if desired.

With regard to the neutral and pleasurable feelings raised by algorithms, the care and support associated with digital services is crucial. Algorithms offer an additional boost to sociality when they identify prospective ties based on a shared location or similar interests, thus helping to form technologically assisted networks, and sort and curate content in ways that feel personal and intimate. As an orienting force of algorithmic culture, the belief that digital services and devices care for us aligns the personal and the political-economic. The convenience offered by algorithmic systems keeps power imbalances in the data-driven economy in place, as it teaches us to coevolve with our algorithmic companions. The following chapter offers examples of how feeling right about devices and services operates as a stabilizing force in algorithmic culture; the perspectives of technology professionals and digital marketers describe how algorithmically mediated practices become embedded in the fabric of everyday life, the guiding idea being that without technologies, we lose something of value. Everyday practices would become lonelier, slower, and less organized without the aid of services that analyze and organize information for us. Think of email without a spam filter, or finding an address without a navigator.

Attending to the neutral and pleasurable feel of algorithms explains why technical systems are felt as inevitable, rendering any kind of resistance marginal or irrelevant. Pleasurable relations with technologies are fed by the anticipatory notion that algorithmic systems are "necessarily coming and therefore always demanding a response" (Adams et al., 2009, p. 249). The present is treated as preparation for a future in which the technologized version of it is what matters most. Rather than aiming for a balanced perspective, algorithmically mediated future developments are seen as particularly valued and worth developing. At times Finnish professionals are very aware of this, and they acknowledge their role in promoting positive and forward-looking algorithmic culture. Through their entrepreneurial attitude, "the future arrives as already formed in the present" (Adams et al., 2009, p. 249). Professionals confirm with their preparatory actions that algorithmic operations align with the everyday in the course of their active adoption and adaptation to their requirements. They knowingly make room for the algorithmic future.

Oppositional Fear

Yet what if things do not improve as relations deepen with machines? The optimistic vision of the greater good that will be achieved with the aid of clever machines is constantly shadowed by the possibility that we are coevolving with digital devices and services in ways that benefit the data-extracting machinery at our expense. The dominant structure of feeling is compromised when algorithmic systems are introduced as matters of convenience but remain matters of concern and, as our interviews indicated, the convenience that algorithmic systems promise is subject to doubt, even among those praising their potential. Hesitations and fears destabilize the idea that algorithmic systems are a source of pleasure, although fear is not an encompassing quality of algorithmic culture or a constant state of affairs; if it were, people would avoid using digital services, perhaps give them up altogether. We are thus not talking about fear as a total social fact or a life-threatening condition. The argument is rather that experiences of fear and discomfort—of not feeling safe—constitute a recognizable realm of the affective infrastructure of algorithmic culture. As a collective experience, fear contributes to an oppositional feeling that counterbalances the pleasures of technologies; it defines algorithms as objects meriting concern. While stimulating algorithmic folklore, fear and doubt operate as destabilizing agents of history, giving form to insecure experiences that outline contemporary digital engagements. Fear, distress, and "mild paranoia" give rise to "a feel" that promotes negative connotations of algorithmic culture.

When combined, articulations of what should be feared present a dystopian outlook on the future, a theme that is familiar from literature and popular culture. Thus the fear of algorithms is not merely born of and maintained by personal experience; public commentators and academics actively feed and nurture the affective infrastructure, highlighting nefarious examples of how technology companies breach notions of privacy and damage public culture. Over the years a number of scandals have drawn attention to troubling sociotechnical developments, with data breaches and hacking events becoming a regular topic in the news. In mid-2013, for example, whistleblower Edward Snowden's revelations about the surveillance practices of national security agencies in the United States were publicized, and his leaked documents received a high level of media attention in Europe. In March 2018 Cambridge Analytica, a data analytics company working for Donald Trump's 2016 election campaign, was reported to have collected and analyzed the personal data of millions of Facebook users for the purposes of targeted political advertising.

Data breaches and scandals are critical events that draw attention to the risks and uncertainties of digital society and raise doubts about its functions. By doing so they mobilize fear and distrust, while experiences of insecurity and loss of control intensify when watching documentaries and reading news articles about them. Popular culture can function as a warning, as in the Netflix series *Black Mirror*, with its focus on dystopian future scenarios. If the pleasures connected to algorithms are imbued with anticipations of a better future, the oppositional feeling resonates with the science fiction thriller movie *Minority Report* (2002), which portrays preemptive technology that facilitates catching criminals before a crime is even committed. In the pessimistic scenario, predictive algorithmic systems capture our thoughts and behaviors in ways that paralyze any notion of free will. Anticipation no longer focuses on the pleasurable but becomes a project of awaiting loss. Whether or not the algorithmic system is able to erase free will is no longer at stake; rather, we need to respond to the felt loss and operate in a crisis mode that assumes that the autonomy to define our own aims has already vanished.

In order to make the oppositional structure of feeling more explicit, chapter 3 engages with experiences of fear and anxiety by applying the framework of the digital geography of fear. On a personal level, feelings of fear and distrust tend to be more intense among those who feel that they lack the agency and skill to master digital technologies. Experientially, the convenient machine turns into a stalking machine, shadowing daily behavior and eavesdropping on conversations. This does not mean, however, that the more technologically competent do not feel insecure and distrustful about current developments. The oppositional feel of algorithms is a combination of things that have gone wrong and that could go wrong. Although data breaches and scandals speak of the same mechanisms of data distribution, everyday concerns remain closer to personally felt worries and insecurities. As Susanna Paasonen (2018b, 215) notes, scandals like that involving Cambridge Analytica are not enough to wipe away the value of Facebook engagements. People still feel good enough about Facebook to use it for their social interactions, which is not to say that they trust it or believe that it is not damaging to individuals or democratic societies. By tracing articulations of apprehension, distress, and insecurity, we begin to see how feelings of fear activate an affective infrastructure that becomes indicative of algorithmic culture. In some ways we never quite attain certainty with regard to all the experiences that contribute to the digital geography of fear, but tracing their articulation aids in trying to capture it. Beginning with the suggestion that the digital geography of fear is a shared experience is an attempt to shift the perspective from the individual to the collective and unpack the patterned nature of algorithmic culture.

Emerging Irritation

Feelings of irritation and frustration illustrate an engaged and indecisive relationship with algorithmic systems, as they are often born of the immediacy of human-machine interaction. As Ytre-Arne and Moe (2020) found when examining how users engage with algorithmic systems—registering their failures, flaws, and imperfections—the Finns who participated in our study did not merely acquiesce or become subservient to algorithmic logics, but critically reacted to what they saw and made sense of it as best they could. People can dream of how to escape the dataveillance and profiling functions of algorithms, only to apprehend how unrealistic it is to imagine they could find a permanent "outside" untouched by processes of datafication.

The emerging structure of feeling is more difficult to flesh out than the dominant and oppositional structures, because it builds on the pleasures and fears related to algorithmic systems but also negates them; indeed, articulations of irritation are in dialogue and tension with them, complicating and complementing both structures of feeling. Whereas the oppositional structure of feeling, comprising fear and distress, introduces a somewhat one-dimensional perspective on algorithmic systems, with the fear sweeping away good experiences with them, irritated reactions underline the presence of both the negative and the positive. In everyday experiences, algorithmic relations are both praised and ostracized, suggesting that tensions with algorithms cannot be fully erased. This means that people have to find ways to tolerate the tensions and live with them. Irritation concretizes how dominant and oppositional forces are simultaneously present when people make sense of algorithmic relations.

The annoyance and frustration offer indications of how such feelings build on and complement pleasures and fears associated with algorithms. Because of the settled and unsettled nature of how people inhabit the digital world, they want to both connect and disconnect (Karppi, 2018). They feel that digital devices enhance their personal autonomy and threaten it. Irritation is triggered when machines do not perform in the desired way, but also by the constant evaluation of whether the digital world serves the needs of individuals and societies, leading to the realization that algorithmic systems can distort our intentions and aims. Flawed or inconsequent algorithms might transgress one's ethical and political convictions, as in the case of Veera, who was unable to block unwelcome fur promotions despite her deep loathing of fur farming. Together the frustrations feed into a broader argument that we should appreciate the irritation as a form-giving social force that is trying to communicate to us what, exactly, has gone wrong in algorithmic culture. As a cultural pattern, irritation might not be fully formed, but if we look carefully at what it does, we can see it as an agent of history that calls for our attention.

In *Ugly Feelings* Sianne Ngai (2005, p. 360) outlines that when we are dealing with feelings, we are not merely analyzing affects or emotions, but "mobilizing an entire register of felt phenomenon in order to expand the existing domain and methods of social critique." The experiences that inform the forthcoming chapters both confirm and reject the fixity and firmness of narratives most readily on offer when it comes to algorithmic relations. Personal experiences are a means to address how top-down story lines of future technologies, but also their critiques, ignore and distort the way algorithms are experienced and lived with. In the current situation, discontents with algorithmic systems do not easily translate into changes in the everyday; despite privacy breaches and informational asymmetries, people rely on digital platforms for a range of routine tasks. Far from demonstrating an uncritical relationship, however, personal experiences propose that people actively explore what might work for them and society. If we cannot detach ourselves from current digital infrastructures, at least we can learn how to distinguish those aspects and features of them that make us most concerned, irritated, and vulnerable. Knowing what causes harm in algorithmic culture, and how we might be harming ourselves in algorithmic relations, paves the way for attempts to move away from the most damaging aspects of technologies and craft more caring responses.

2   *Coevolving with Algorithms*

It is not difficult to think of ways that digital services improve the everyday. The participants in our study listed them with little effort: Google Maps provide a pleasurable way of navigating the world; social benefits can be accessed without having to send hard copies of documents by mail; if you have children abroad, you can communicate with them through Skype; Facebook groups for secondhand goods are convenient for making inexpensive purchases; digital clinics offer advice without a physical visit to the doctor's office; and if you have a rare disease, you can find fellow patients from different parts of the world to share your experience with. Digital services offer a helping hand in the everyday, as they erase physical distance and open new venues for communication. Among our educated interviewees, algorithms are largely thought to offer assistance in personal, organizational, and societal matters, as they allow them to extend the range of human capabilities and agencies.

The pleasures associated with algorithms are further enriched by active appropriation. In this, however, the interviewees differ from each other, as some dedicate time and enthusiasm to technology relations while others do not. Liisa, a nutritionist in her fifties, treats algorithms like the weather. She takes them for granted, and she believes there is little that she can do to change the state of affairs. To mobilize the productive capacities of algorithms, people need to understand them well enough to influence them, even if this is not consciously articulated or acknowledged as a skill (Baym, 2013). As Sarah Pink and her colleagues (2018, p. 3) claim about the ability to act: "It need not involve absolute certainty, but entails feeling and knowing enough to be able to take the next step."

Oskar, who has a vocational qualification in business information and publishes vlogs and music, is a passionate proponent of digital technology. He treats technological evolution as an extension of human evolution, praising the vision that algorithmic systems offer for the future, as machines become our guides and companions. Oskar describes how AI has defeated human players in chess and the game of Go—a strategic board game invented in China over two millennia ago—and helps in providing solutions for complex issues like climate change and resource allocation that a human cannot even imagine. Since biological and cognitive evolution is so slow, AI introduces new kinds of prospects for progress. Oskar feels, rather than knows, that an algorithmic system is an enabler that allows humans to better exercise their often limited capabilities. He says that large-scale systemic problems, including climate crises and global poverty, become solvable as data-driven technologies expand the range and scope of human imaginings.

This chapter picks up the question of how the pleasurable realm of the affective infrastructure of algorithmic culture comes into being in the everyday. It attends to the dominant feel of algorithms by way of personal experiences, to demonstrate the care and support associated with algorithmic services. The examples of how devices and services "feel right" explain how technical systems become embedded in mundane practices and how they are seen as important preparation for the future. When algorithms become participants in daily lives, they deepen the collaboration between humans and machines. Attending to coevolutionary processes allows the exploration of how technologies are envisioned as active companions, suggesting an unbroken association of the human and the algorithmic system (Kristensen & Ruckenstein, 2018). By following what is said about algorithms and feelings connected to them, we begin to get a better sense of what is—and is not—implicated in processes of coevolution. The coming together of humans and machines takes place in many different ways, and not without tensions. There is always the possibility that the coevolving is distracting us from what is important to us in life. Yet anticipatory and pleasurable engagements with algorithmic processes continue to strengthen the coevolving of humans and technological companions. The pleasures articulated in relation to algorithmic techniques underline their agentic qualities, as well as the fun of learning and experimenting with them.

The Holy Computer

Henrik, a life coach, explains that he celebrates the hippie movement and living in harmony with nature but still thinks that the computer is the key to the future of humankind. His thoughts are influenced by the techno-libertarian ideology set in motion by celebrated forces of the cyber culture world, including Stewart Brand, Kevin Kelly, and Nicolas Negroponte (Frau-Meigs, 2000; Turner, 2006). Since the late 1960s Brand, and others associated with Whole Earth publications, had been linking information technology to "New Communalist politics of personal and collective liberation"; this work was seen as an extension of the 1960s' consciousness movement (Turner, 2006). The new digital generation set out to dismantle hierarchies; they wanted to destabilize dominant corporations and governments and create a collaborative society, interlinked by currents of information. Drawing inspiration from the ways computational forces can be harnessed to shape awareness and "make the world a better place," as the Silicon Valley mantra goes, the Finnish life coach says that he sinks into deep reflection when pondering how humanity could progress and flourish when shaped by digital technologies. The techno-libertarian spirit that moves across localities from Silicon Valley to Helsinki offices begs the question of how one can—personally—become part of the global stream of humanity and assist in the task of reshaping the world.

Henrik mentions that some people he encounters deprecate his technologically inspired vision and what he most values—digital technology, marketing, and entrepreneurship—as superficial and phony activities, of little worth and possibly even dangerous to humanity. He strongly disagrees with the critique. He treats algorithms, particularly those used for pattern recognition, as detection mechanisms that allow us to engage with the unknown worlds of present and future. Here, algorithms are associated with their proficiency in offering visibility, the idea being that by making previously unknown aspects of life detectable, we can gain more control over them. As Joseph Davis and Paul Scherz (2019, p. xxxiv) say of this kind of thinking, "The world is a mathematical puzzle, and quantification is both the way to understand it and the means to solve it." For Henrik, algorithms are a way to discern what humans fail to discover and to penetrate the "code of the world" in order to understand what humanity is truly about. In this framing, algorithms become associated with time-honored tropes of conquering previously unexplored territories and making them available for mapping and tracking (Edwards et al., 2010; Haraway, 1998). Henrik's religion-like conviction that if we embrace algorithmic potentialities and orient ourselves toward them, great things will happen, assigns the computer an almost holy status in the sense that with the aid of algorithms it can do unexpected and perhaps even magical things by detecting patterns in the world. The enchantment and magic of technologies, in this case algorithms, is sustained by hiding the human labor involved (Suchman, 2007). Providers of AI services can strategically promote the idea of machinic self-sufficiency by obscuring and occluding human efforts that go into the design and implementation of such services (Newlands, 2021). The dominant structure of feeling validates the capacity of algorithms—and algorithms alone—to open new paths and trails with their seeing and knowing capabilities.

Best of Humans and Machines

Pauli, a lecturer in business, describes how computational intelligence improves communication and in doing so, shapes selves and socialities. Algorithmic predictions trigger lively speculation about what more the machines could do for us, promoting the anticipatory notion that algorithmic systems are not only valued but worth developing. Think of accounting, for instance. Instead of the accountant checking a thousand invoices, he examines only a hundred that the algorithmic model has picked up due to unique or suspicious features, which require more careful human review. Similarly, in retail, workers' schedules could be compiled with the aid of machine learning: the data collected from the grocery store's checkout registers, human traffic flows, and advertising, which is likely to increase the number of clients, could be combined with data that details the wishes of employees concerning their workday rosters. With mind-numbing and tedious tasks like producing shift lists delegated to machines, humans can oversee larger organizational matters and perform more rewarding tasks. Rather than working like a machine, or doing chores for a machine, training and supervising algorithmic systems gives people the upper hand.

Pauli predicts that algorithms will be everywhere, promoting various kinds of data relations from the mundane to the infrastructural. In the mundane register, the office coffee maker will identify users by their thumbprint and automatically serve their favorite coffee. Infrastructural data relations, on the other hand, speed up the distribution of services in a societally beneficial manner. Rather than making food-delivery couriers work in precarious conditions as low-paid "partners," couriers do not need to be human at all. Robots and autonomous vehicles can deliver food from restaurants to customers. Pauli imagines how systems that have been developed in commercial contexts could be used to strengthen social ties and networks at work and in neighborhoods. The matching functions, familiar from dating applications, could be applied to other contexts as well: finding like-minded colleagues outside of the workplace and building safety nets against the precarity of work life. Lonely people could find company in their neighborhood.

Oskar, publisher of vlogs and music, however, says that realizing the potential of technologies entails that not only machines, but also humans, evolve, a theme that comes up repeatedly when professionals point out that the implementation of technologies that would aid human progress remains challenging, if not impossible, without the expertise to steer developments in the right direction. Technologies offer possibilities for organizational growth and large-scale societal modifications, but realizing them requires will and work. What Oskar and like-minded professionals argue is that only by carefully combining human strengths and machine strengths can machines truly serve us. Here, their thinking aligns with Joseph Aoun's *Robot-Proof* (2017), in which he argues that we need "to collaborate with other people and machines while accentuating the strengths of both" (Aoun, 2017, p. xx; see also Schüll, 2019). Computers might surpass us in cognition and precision, but they do not have the capacity to relate to another human's feelings, concerns, or inspiration.

A powerful vision of the algorithmic age is that humans become more human with the aid of machines. Aoun praises the ability of humans to craft imaginary stories, invent works of art, and construct notions that illuminate reality (2017, p. 21): "Only human beings can look at the moon and see a goddess, or step on it and say we are taking a leap for all mankind." He proposes that working closely with machinic agencies will teach us to value the uniquely human: our ability to feel, socialize, improvise, and make sense of life. The dominant feel of algorithms is maintained with associating humanizing rather than dehumanizing forces to machines. While algorithmic systems perform computational functions, people can focus on empathetically connecting with one another, experimenting together, and building caring relations. The flipside, suggesting instead that we are adjusting our actions to machines, following predefined rules and procedures, and becoming more machine-like in our behavior—developing into human algorithms—is downplayed by the emphasis on increased agency and possibilities for creativity.

Everyday Automation

Everyday techno-relations emphasize that algorithmic culture is not merely fed by what people think, advance, or anticipate, but is firmly rooted in what people feel and do. While contemporary relations with technology might be promoted with optimistic and forward-looking conviction by some, coevolving with digital technologies is typically more unexciting and routine-like. In thinking about habit formation, Annette Markham (2021, p. 388) describes how she uploaded a photo to some "sort of cloud storage"—most of us do not really know where the pictures go—and created "a digital pathway" that did not previously exist. Over time, Markham reflects, that pathway became a habitual way of storing photos, until the pathway itself disappeared. She no longer thought about the process or how the system learned preferred actions. The uploading of photos became habitual. There was no choice involved; the act of doing it just happened. The human agency does not disappear, but it is "buried beneath the seamless accomplishment of a goal" (Markham 2021, p. 388), and the digital pathway defines action and direction.

In light of such habitual pathways, it is not surprising that it is common to discuss algorithms as if they were neutral instruments to get things done, as Heidi, who studies at a business school, explains: "There is no reason to demonize algorithms, because they make our lives easier." Cecilia, in her twenties, studying media in a polytechnic and working for a broadcasting company, thinks that the term "digital native" aptly references the role of computers in her life since primary school. She cannot think of her everyday life in a computer-free register, regarding the generational gap as the main experiential digital divide. Younger people are inescapably tied to mobile phones, their new "companion species" (Lupton, 2016c; Rettberg, 2018), which act as personal archives and memory extensions; phones can be discussed in much the same way as pets and soulmates. Like Cecilia, others in their twenties talk about how older generations can function socially without being connected through apps; they appear to have more life outside the digital and less fear of missing out. They still contact each other with regular, old-fashioned phone calls that youngsters avoid. Young people have grown up with technically mediated sociality—programmed sociality, as Bucher (2013) calls it—without any conscious thought about whether it is good for them or for society more broadly.

When people are first introduced to digital devices and services, they tend to use them in a wary and conscious manner. Once services are fully adopted, however, they become an integral part of how everyday lives and worlds are experienced. Selma, who has an arts degree, reflects that coevolving with digital technologies means that people are less concerned about privacy violations, for instance. She is not troubled by this because she cannot imagine people being taken advantage of by data companies on a massive scale. "Perhaps I am naïve," she says, "but it is hard to believe they would want to do anything evil. Of course they want to benefit, but do I lose something in that process?" The invitation to reflect on what is at stake in algorithmic relations takes Selma onto uncertain terrain. The dominant structure of feeling, strengthened by the convenience that technologies offer, is inescapably tied to market forces and the associated quest for profit. Sensors and communication technologies inhabit the everyday, as they extract value from data traces, while in exchange for such traces users of services are offered a foundation for their sociality, including guidance on potential places to go and people to meet. Selma is not alone in doubting whether she loses something when commercial advantage is gained from data about her behavior, because it is difficult to think of cost when nothing tangible is being lost. Algorithmic relations are characterized by an invisibility that makes it difficult to observe who is exploiting whom, and based on what criteria (Paasonen, 2018b).

On social media, small acts—how we read, watch, scroll, or click—influence algorithms. Everything we do is potentially a source of digital data, and findings based on that data might be fed back to us. Yet people keep doing things with little thought of how their activities relate to apps or algorithmic operations. Some say that they are on social media platforms all the time; others report that they hardly use them but still constantly check their accounts, a practice that has become so quick and habitual that it is no longer registered. Users might blame the algorithm for irrelevant content without realizing that their own behavior contributes to the way the system performs (Schwartz & Mahnke, 2020, 9). As algorithms mirror everyday behavior, they will tot up and respond to the hours spent watching funny cat videos or reading sensationalist stories, even by those who regard themselves as fact-driven media consumers. From this perspective, not only are we data (Cheney-Lippold, 2017); our actions are also present in algorithms. Algorithms begin to resemble us, as they generate and package information about life processes—about what we do—and use our behavioral traces to offer us predictions, recommendations, and valuations.

Algorithmic Care

Maria, a technology communication consultant in her fifties, is pleased with Oura, a self-tracking device designed as a ring that measures heart rate variability and condenses the analyzed results into three simple scores: sleep, readiness, and activity. Algorithmic care is designed into self-tracking devices in order to assist and reinforce wellness aims (Schüll, 2016). For Maria and many others, frictionless co-living with algorithmic systems is pleasurable and rewarding. Here, the users and designers of services are on the same page: they are both after successful co-living with computational tools. Maria describes how she optimizes her sleep with the goal of being at her sharpest at work. If she is readying herself for a particularly demanding assignment, she starts her preparations days ahead, making sure that she does not eat anything heavy that could weaken the quality of sleep. She avoids alcohol and goes to bed when she feels tired and not according to the clock. In the optimization task, she uses Oura to observe her bodily signals. For Maria, the Oura ring is an integral part of her excellence pursuit, as her desires and the communicative qualities of the technology operate in perfect balance. She nurtures her heart rate variability with the aid of Oura to have "all her receptors ready, brain cleansed, and heart beat balanced." If all is good, the Oura ring confirms her readiness and tells her to go for it with the catchphrase "carpe diem."

In order for the algorithmic relations to feel gratifying, people need to feel that they are in charge. The tracking and measuring of lived lives with smartphones and watches develops into a caring practice when people rely on their devices to check whether the calories burned, steps taken, and hours slept support their attempts to lead balanced everyday lives. Thus in the positive register, a self-tracking device that announces that it is time to go for a walk or seize the day offers much-needed guidance. With the aid of sensors and devices, people try to eliminate disturbing practices—eating too much, wasting time and not being productive enough at work—and develop better ones. If personal goals align, at least initially, with the goals promoted by the device, the partial or temporary loss of self-directed action feels rewarding. As Jeannette Pols and others argue, this kind of alignment is crucial for well-adjusted engagement with devices, as it provides tools for "self-induced nudging into self-prioritised activities" (2019, p. 101).

When balancing care with autonomy pursuits is successful, seamless collaboration grows into a process of coevolving, with action and intentionality resulting from the interactions of the human and the machine (Kristensen & Ruckenstein, 2018). Technologies assist in increasing the consciousness of one's agentic capabilities and heighten awareness of mundane everyday doings, suggesting that self-tracking might be a way to learn what one already knows in an embodied and unreflective manner (Fors et al., 2020). Devices like the Oura ring offer sensing support, but they might also engender sensory experiences. In their ethnography of hypoglycemia, Annemarie Mol and John Law (2004, p. 48) discuss how the use of blood glucose measurement devices "train[s] inner sensitivity," promoting what they call "intro-sensing." Gina Neff and Dawn Nafus (2016, p. 75) argue that personal data can "become a 'prosthetic of feeling,' something to help us sense our bodies or the world around us." Sarah Pink and Vaike Fors (2017, p. 376) note that self-tracking mediates "people's tacit ways of being in the world," promoting an awareness of mind-body in the environment. The increased responsiveness in terms of sensory impact—the felt effects of eating, stress, or exercise—generates "sharpening of the senses and the "production of new senses" (Kristensen & Ruckenstein, 2018). At the core of successful algorithmic relations is the alignment of interests and the balancing of practices of care, as people give away some of their self-determination to a technical system. Devices that operate at the intersections of the body and daily lives are too diverse to suggest a shared or identical response. What is crucial, however, is that they promote new patterns of experience. The Oura ring has taught Maria to think differently about her sleep and act accordingly. While she is living the sleep metrics, she becomes a prolific and skilled sleeper.

The Machine Knows *Me*

Henna, a student of theoretical philosophy, describes the convenience of finding inspiring content with the aid of algorithms. If she gets excited about a genre of music or fashion, she can easily find related materials and tips. Algorithms speed up the circulation of relevant information and pleasantly surprise her with timely product recommendations and notification of stimulating social events. Recommender systems become companions in the everyday as they shape practices connected with finding dating partners, music, and movies.

Commercial agents downplay the difference between commercial and noncommercial relations with the rhetoric of sharing and participation; for example, "the nearby event" advertised on Facebook might be arranged either by a group of friends or a profit-making enterprise. By emphasizing convenience rather than their concerns, the respondents of our study highlight how important it is for the encouragement to consume to align with their interests. Advertisements that hit the target trigger an enjoyable sense of being recognized, a feeling that *the machine really knows me*. Erika, a chef, is one of many who enjoy ads being timely and suitably personalized. She laughs about a top that she "had to get" that had exactly the right slogan for her: "I'm not running for fitness," it proclaimed, "I'm running because they are taking the hobbits to Isengard." The pleasures gained from algorithmic techniques include intimations of care and recognition when they boost the experience of being seen. Erika is not merely being sold any old top; the *Lord of the Rings* reference responds to her idea of how she would like to be seen by others. The sensation that the machine knows the user energizes and upholds pleasurable algorithmic relations.

Algorithmic techniques draw companies away from traditional advertising models toward digital marketing that dynamically sorts and sells data traces to generate marketing that feels relevant. "The new ideal is a personalized presence that is so embedded in daily routines that it becomes second nature" (Fourcade & Healy, 2017, 23). Erika experiences this personalized presence when she purchases the top with the hobbit slogan. Yet the personalized presence is also a surveilled presence, a fact of which Erika is acutely aware. "I am being watched," she says. A dialectic consisting of the dominant structure of feeling—activating feelings of being cared for and pleasantly recognized—and oppositional forces comprising disapproval and suspicion is in play when people evaluate algorithmic relations. Articulations of emotions oscillate between the pleasures and fears triggered by living alongside and being enmeshed with technological systems. It should not be overlooked, then, that the convenience that algorithms offer is an ongoing source of cultural tension as it supports the commercial co-optation of everyday aims and practices. The intimacy of the relationship speaks to the ambivalence of pleasurable personalization and unpleasant surveillance, underlining their concurrent presence in algorithmic relations. Pavlos, a machine installer who has moved to Finland from Greece, wishes that service providers would become a bit more relaxed in their data-gathering efforts, grumbling, "They don't need to know everything that I wish from life."

Algorithmic matching remains a delicate balancing act, with marketing operating on the borders of personal and societal tolerance. In order to promote coevolving, the goal of designers and digital marketers is to be intrusive, but not so intrusive as to offend. To reach their goals, professionals try to gently break the resistance that stops consumers from clicking on advertisements and purchasing products. They seduce people into actions by offering nudges and baits. In this sense, all digital marketers are growth hackers, aiming to shorten the time that it takes for people to click on advertisements and purchase products. Rafael, who has worked in social media marketing but is currently unemployed, notes that the better the algorithms do their job, the more efficient they are in hooking people, thereby creating virtuous cycles that keep advertisers happy and the social media company thriving. Yet the downward spiral is also difficult to break. Intrusive ads annoy users who then use the service less, giving the advertisers less visibility. In practice, this means that users should not be pushed too hard, as the gratifying coevolution with algorithmic systems can quickly become an appalled and resistant reaction.

Henna, the student of theoretical philosophy, is content with the algorithmic supervision that she gets, but she also wonders whether she sees all the important content. The personal guidance offered by algorithms raises the question of what exactly is appropriate and for whom. Perhaps the selection process is guiding her in an unfavorable direction? For Rosa, the most alarming aspect is that people are being unconsciously guided into certain modes of thinking. She represents the many when she wonders whether

her opinions are still hers, and whether she can still think independently, with her own brain. Her questions resonate with the felt loss of a "breathing space" in which to reflexively develop a sense of self in the midst of data-extracting technologies. Professional resources, taking advantage of the tests and tools of behavioral psychology and economics and mobilized to push against consumer resistance and autonomy, require a new kind of attentiveness to what one wants and feels. Iida, a concierge and part-time student, describes the risk that she will end up purchasing things that she does not need or already has. What she appears to be saying is that the machine should know me, but not in a manner that would jeopardize my personal autonomy.

Shaping Digital Spaces

Anne, the high school teacher, describes how she anticipates, even before she clicks, what will result from her actions. Her algorithmic engagements could be described as preemptive and deliberate. She "clicks consciously" (Bucher, 2018, p. 109) to instruct the algorithms about the kind of information she wants. If she is searching for kitchen shelves online, she thinks twice before she clicks on Facebook ads, as this would send her down the algorithmic rabbit hole: her feed would be flooded with furniture ads for weeks to come. She adds that a colleague, a runner, uses incognito mode when buying sneakers, illustrating the impact of targeted ads with an offline equivalent: how would it be if, after visiting a shoe store, the salesperson, and then salespersons from other stores, tirelessly followed him around pestering him with different sneaker models and even matching T-shirts and pants? Anne had never thought about targeted advertising this way, but says that it is precisely what we should do: think more with offline parallels to highlight the distorted nature of the digital world.

"As machines merely process information they have, I need to signal to them what I want," Kasper, an editorial assistant, explains. He clicks on posts or news items to communicate their worth to the algorithm. His goal is gradually to teach the machine the kinds of information he values, and to this end he offers behavioral clues and indications, material to which the algorithmic system can respond. This kind of human-machine interaction underlines that gratifying experiences with machines can involve considered engagement, small everyday acts that try to influence how algorithms behave. Mikael, an anthropology student, tells us that he acknowledges all the posts of Antroblogi, a Finnish online blogging community, to aid the circulation of its content. He adds that it is better to react with an emoji, rather than merely a like, because reacting is supposed to give more visibility to a post. Nobody has taught him that "this is how you influence algorithms," but it feels like the natural thing to do. Without thinking much about it, he has internalized "the popularity contest" (Bucher, 2018, p. 105) as a defining feature of how algorithms sort and rank content. Mikael's observations tell how "influencer practices" take place beyond the advertising and marketing profession; online we can all become everyday marketers, promoting ourselves and our causes as influencers. The market is teaching us to see ourselves not only as data-generating subjects for the corporate surveillance machinery but also as everyday marketers, capable of shaping the world with the content that we have chosen.

In a darker register, the popularity contest could be called the tyranny of the algorithm, as the same influential people and the same type of promoted content are highly evaluated time and again. Yet the structure of feeling that resonates with the convenience and pleasure of algorithmic engagements tends to decenter questions that have to do with how the platforms' business model values paid content over unpaid; instead, it emphasizes that the social power of algorithms stems from "recursive relations between people and algorithms" (Bucher, 2018, p. 116). According to the dominant structure of feeling, it is not only the platform that operates as a gatekeeper and decides what the algorithmic space looks like, but also those who like, share, and comment. Mia, a sociology student with computational skills, says that she feels that she is unaffected by algorithms, an opinion also shared by others; people think that they can influence social media content without being "determined" by it. The acts of clicking sustain a sense of ownership and autonomy, no matter how factual such influence is. Algorithmic mechanisms are treated as material to work with; if algorithms are persuading us, we can persuade them back.

Henrik has put a lot of energy into making his Facebook page—which is what he sees when he logs onto the site—"a nice place to be." The inspiration for this effort was triggered by the heated controversy over immigration, locally termed "the refugee crisis," that began in Finland in 2015 with the advent of unusually large numbers of asylum seekers. Refugee-related discussions were already characterized by racist overtones prior to the crisis, but the crisis temporarily escalated patriotic and nationalistic themes on social media to an unprecedented degree (Nikunen, 2015; Pantti et al., 2019; Pöyhtäri et al., 2019; Ylä-Anttila, 2020). With the news of the growing numbers of refugees, Henrik felt that his social media newsfeed had turned into a minefield; every morning he anticipated with some distress the negative posts that would ruin his mood. Since he wanted to avoid distractions at work, he actively manipulated algorithms to reflect his aims. He selected what he liked, clicked, and consciously avoided; he blocked twelve hundred of his fifteen hundred Facebook friends because their updates were discouraging. His Facebook page is wonderful these days, he says, as it only contains optimistic stories about technology, animal videos, and dog memes. Here, algorithms are an integral part of securing an elevating emotional tone and erasing content that could sidetrack him. Through his content curation, Henrik has aligned his Facebook page with the dominant structure of feeling, featuring technologies as optimistic and forward looking.

Laura, a photographer and a self-proclaimed feminist, has more politically resistant aims in mind. She jokes that she would like the algorithmic system to learn that she wants only female-dominated content, that any algorithm that influences her, or shapes her life, should perform according to her ideals and aims. Consequently, she meticulously purifies her social media newsfeeds of posts and advertisements that reflect a patriarchal world order, working as an everyday content moderator by cleaning the algorithmic space to reflect her values. Whereas commercial content moderators guard the limits of conversations and clean the digital space of content seen as "dirt" (Roberts, 2019; Ruckenstein & Turunen, 2020), everyday content moderators, like Henrik and Laura, not only delete unwanted content but also purposefully shape the digital space around themes of their own choosing. Henrik seeks to build a techno-optimistic fun sphere, while Laura's goal is to create a politically safe space by combating patriarchal influences. As algorithms operate in contexts defined by existing social stratification and related inequalities, they will continue to reiterate social divisions and gender hierarchies unless they are subjected to interventions in a sustained and proactive manner. One individual will not change the state of affairs, yet although the patriarchal world order cannot be erased with a few clicks, it remains important for Laura to envision change and act accordingly—if for no other reason than to keep in mind the feminist goal that things could be otherwise.

Anticipating Perfection

The examples so far have suggested that algorithmic relationalities have different qualities depending on how their persuasive powers are responded to. When experiences of targeted advertising are compared with recommender systems in subscription services—in our material mostly Spotify and Netflix—further light is thrown on what feels right in algorithmic relations. Cecilia thinks that it is only logical that targeted advertising irritates, because it is unsolicited advice, random persuasion. She exemplifies her irritation with a reaction to a frequent commercial on YouTube featuring crying children: "Well, no, I am not going to have children for your sake." Unlike targeted advertisements, Spotify offers Cecilia inspiration by recommending music that she likes. She says that she willingly submits to the tracking of her behavior and the conversion of her consumption of music into lists, charts, and recommendations. Since the recommender system has learned her taste in music, it pleasantly surprises her with suggestions and introduces her to bands that she would not learn about otherwise. Her music consumption starts to take on qualities of coevolving, as she is no longer solely in charge of what she chooses to listen to.

Cecilia is describing a relationship to Spotify that is shared by many. By clicking purposefully and engaging in "feedback-giving practices" (Siles et al., 2020, p. 12), Spotify users aid the recommender system and make their desires explicit. They follow and like songs and artists, or skip songs they detest, and repeatedly listen to their most preferred picks. They would probably repeat listen anyway, as people tend to play their favorite songs time and again, but telling about the activity accentuates that the transfer of taste in music to the algorithmic system requires ongoing signaling. Only by sustaining continuous communication can the recommender system develop into a companion able to predict what one wants. The promissory qualities of the recommender system are heightened by the close relationship, built on data transfer, enabling human-machine coevolution. In the process, the collaborative relationship gradually builds the expectation that the algorithmic buddy will react to the information it is fed and improve recommendations. The recommender system not only aids in the discovery of new music but participates in revising and modifying ways of listening to it (Karakayali et al., 2018). Similarly to a pedometer that becomes an integral part of renewed walking practices, the recommender system suggests new ways of listening to songs, revising habits of music enjoyment.

Cecilia ponders whether the responsiveness of the system teaches her to expect more from it. The desire for upgraded service, however, is not a conscious request for better-targeted content, but rather a side effect of coevolving: a quiet but persistent demand that algorithms should respond more intuitively to her doings. Here, the convenience that the algorithmic system provides is not a constant quality; rather, it is attended by ever-growing expectations that have to do with the nature of the human-machine relationship. As the human is coevolving with the machine, the relationship needs to develop and mature accordingly. When a certain level of performance is achieved, more input is expected.

Human-machine connection thus takes on an anticipatory outlook characterized by a yearning for flawless compatibility.

The anticipatory attitude calls for the algorithm to fulfill its part of the mutual relationship (Siles et al., 2020, p. 12). With accumulating data traces and a relationship of coevolving, it is only logical that the recommender system should improve in responding to personal expectations; if it does not, it can generate feelings of discontent. Jasper, a PhD student in sociology, describes how he has tried to influence Spotify's recommendations by actively upvoting and downvoting suggestions to improve the system, but his actions have had little effect. Considering the volume of information that he has readily offered about his taste in music, he finds it surprising that the system suggests "the same lame pieces" that he has already rejected. He treats the nonresponsiveness as "a fatal flaw" in the recommender system. Luckily, he adds, Spotify has introduced a special button for disqualifying certain artists altogether, which he thinks will help, even if it does not solve the problem altogether.

Companies frequently add new features to recommender systems to improve them. The development of such systems is endless, at least in theory, while adding complexity to how content is combined, sorted, and ranked presents new tensions for human-technology relations. Yet optimistically, there is always the possibility that the system will be amended to respond in a more proactive manner. The anticipatory orientation in human-machine relationships sustains the dominant structure of feeling, while the longing for the perfect algorithm becomes an integral part of the everyday pleasures of coevolving with algorithmic systems. Based on what it should do, the perfect algorithm is active, offering personalized and preemptive action and variety, surprises, and nonstereotypical suggestions. Ironically, a truly pleasing algorithm is no longer very machine-like at all. In taking stock of human desires, the perfect algorithm internalizes them so efficiently that it becomes like the human that it shadows: intuitive, context aware, and responsive to change and diversity. Anticipating perfection from the algorithms, then, generates a longing for a more human approach, a theme that is also present in the irritation and frustration that shape the emerging structure of feeling in algorithmic culture (see chapter 4). The dominant feel of algorithms is strengthened by promises of technologies becoming more like humans, yet when humans engage with technologies, their nonhuman nature is also obvious, triggering the hope that to assist the algorithm, there should be a way to communicate what is important and worth promoting.

Experimenting with Dividuals

The examples that feature relations to recommender systems have described how the feedback loops tighten the coevolving of humans and their algorithmic companions. The pleasures of algorithms, however, can also be linked to practices that try to actively create feedback loops that would promote desired results. Technologized futures are promoted with an exploratory stance. Tom, the part-time event planner, praises all the things that he can do in the digital marketing realm with the aid of algorithmic techniques. For digital marketers, the possibility of measuring the reach of their campaigns and observing attention and human behavior more broadly adds a new kind of excitement to their work. The optimistic and forward-looking stance in relation to algorithmic operations has to do with the role that digital marketers play in trying to find insights into human wants, needs, and goals with the aid of data, then acting on the insights through experimentation.

The enthusiastic stories of Tom and other digital marketers describe how learning and experimenting with algorithmic techniques promote processes of coevolution. Here we are at the heart of the friction approach, which pays attention to how the processes of datafication become personally felt invitations to participate in market developments. The goal of data gathering and tracking is to intensify a phenomenon, not only to measure or predict it but also to engage with and examine it. Tom works with algorithmic techniques, affirms them with his practices, and by doing so paves the way for marketing techniques that shift the perspective to how people's lives are perceived. He experiments with computational features: the object is the dividual rather than the individual. As the data resources need to be compressed into a legible, easily approached format, computationally skilled digital marketers carry out their work by means of data analytics and data visualizations, suggesting that these methods produce insights that benefit the planning of marketing operations (Pääkkönen et al., 2020).

Administrators of social media campaigns can see at a glance in nearly real time which market segments like certain posts and how people interact with ads that are posted at specific times of the day or week. Tom relates that if the data analytics suggest that performance is not as good as it should be, he updates his event campaigns around trying to understand how to increase user engagement, working with an iterative logic to design improved operations. From his perspective, digital marketing is more interested in tools and measurements, experiments, and rounds of iterations than in the users of digital services targeted by the campaigns. This explains why digital marketers downplay the fact that they are handling personal data; they think of the data as dissociated from the lived lives of individuals. As Tom argues, "Marketers are not interested in anybody's private life. They are interested in, 'I got my message through and that message is working.'" Instead of people's behavior, digital marketers might emphasize that they are tracking their own performance based on data such as how many views, likes, comments, or shares a post receives. Only certain datafied aspects of life are of interest to them: life events, for instance, or characteristics, hobbies, daily routes, and travel. The machine-readable situations, including pregnancy, divorce, renovation, age, gender, and holiday, are valuable as they can suggest what might appeal in terms of purchase decisions. The reach of a campaign translates into numbers, and once the numbers attain a preset goal, the campaign is regarded as successful (Kennedy, 2018, p. 23). Experienced digital marketers might be interested in reflecting on what lies behind the numbers, rather than in taking them at face value, but for their clients and less qualified marketers, numbers can become an aim in themselves: inaccurate data and results are accepted, as long the "desire for numbers is fulfilled" (Kennedy, 2018, p. 23). What the likes and reactions signify, or how the scores used to evaluate their performance are calculated, lose their importance if the numbers are high, indicating bigger audiences and additional user engagement.

Self-taught digital marketers can use digital tools in an ad hoc manner and admit to guessing and experimenting with what might work in the algorithmic environment: a game-like feel to such efforts can make them fun and, when successful, deeply gratifying (Cotter, 2019). Yet enthusiastic engagement with data analytics does not always translate into great algorithmic skills. Joose, who promotes his music online, relates how he accidentally clicked something on Facebook, thereby purchasing visibility. He only discovered this a couple of months later when he received notification of an unpaid bill. In the meantime, Joose had detected a curious detail: almost all the likes on his post were from Indonesia or other developing countries. This made him speculate whether, once you purchase visibility, the company providing it merely buys likes from a click farm. If the visibility has not been specified in any way, it could be anything, including meaningless reactions from click workers in faraway countries.

Ultimately, getting your messages through on Facebook or Google is a question of money. The more you pay, the more visibility you get, although some digital marketers are convinced that it is possible to succeed with creative and clever campaigns. The pleasures of marketing align with those of biohacking, with the difference that here the exploration is not geared toward one's own body and mind, but toward the behavior of unknown others (Ruckenstein & Pantzar, 2017). Digital marketers use digital traces as material for their tests and trials when trying to steer people's attention and behavior. Thus, experimenting with algorithmic techniques is not merely about the application of tools; it changes our perspective on ourselves and the people around us. Digital marketers enjoy algorithmic techniques, while they apply machinic categories to human processes at an accelerating pace. Pleasures of algorithms become connected with the social inequality of our times, as technical qualities determine how other humans are handled. When digital marketers approach consumers as dividuals, they deconstruct them and treat them as data sets without human qualities. At the same time that digital marketers enhance their own agency with technical aids, they reduce the agency of others by treating them as bundles of computational traits. The algorithmic age becomes visible as a tendency to dehumanize selectively. Those who can dehumanize others are on top of technologized futures.

Maintaining the Flow

This chapter has argued that the dominant structure of feeling aids in the expansion of digital services in people's lives, as it promotes projects of how we should work and entertain ourselves in the algorithmic era. The structure of feeling becomes observable with the aid of personal reflections that constitute an undercurrent of technology relations, keeping contemporary power relations and informational asymmetries in place. The current tech landscape is built on existing social stratification, possibly deepening inequalities with the tendency to dehumanize those who are not part of defining technologized futures. Yet the neutral and pleasurable feel of algorithms breeds observations of algorithmic convenience, strengthening historically rooted notions of technologies being good for us. The pleasurable feel of algorithms not only sets the tone for technology relations; it keeps them going. The dominant structure of feeling relies on a forward movement, opening rather than closing of options. In light of positive assessments, people are not being defined by algorithmic techniques, but they are learning, growing, and evolving with them. They try out applications and devices, acquiring new knowledge in the processes of familiarization. From this perspective, algorithmic techniques are not interpreted as restrictive and reductive; rather, they open ways to discover insights, improve existing practices, and promote new ones.

Active engagement with technologies deepens commercial involvements as it invites devices and services to become participants in the everyday, yet mastering those involvements is also a way to limit uninvited commercial co-option. Professionals in particular stress the many possible directions for market relations; there is always the opportunity to contest and develop something new. "Markets are contingent," Liz McFall and others argue (2017, p. 14), "upon the associated action of individuals in attaching, rejecting, complaining, negotiating, reviewing, modifying, hacking, appropriating and refusing market offerings." While commodification is a process that seeks intimacies and alliances in the interrelationality of humans and technologies and parasitically latches onto them, people can ignore and avoid such unity of aims and adapt technologies to other ends and alternatives. Consumers are neither entirely free to make their own choices nor victims of the market; rather, they are complicit in promoting new ways of living with algorithms. Not coincidentally, those most eager to endorse algorithmic futures are ready to learn new skills and promote practical and communicative engagements that generate conditions for current and renewed practices. Algorithmic market-making efforts depend on identifying and attracting people who have the time, patience, and enthusiasm to become involved, although the active translation and contextualization work conducted by users of products and services often goes unnoticed (Pinch & Oudshoorn, 2005).

Engaging with the dominant structure of feeling underlines the importance of staying attentive to the emergent and transforming area of communicative and agentic relations between humans and machines. Pleasurable cohabitation with algorithmic systems strengthens the reach of algorithmic culture by naturalizing technology uses. By bringing to the fore the very intimate ways in which algorithms become a part of people's relationship with themselves and others, we can observe how coevolving with machines aligns with the machinery of global data extraction. In order to become universally appreciated, algorithmic techniques and related concepts and ideas need to travel across differences. Everyday experiences involve ongoing assessment of whether algorithms' convenience—their ability to smoothly respond to personal or societal desires—starts to harm our self-understandings, social relations, and societies. While boosting productivity at work by means of sleep tracking seems like a living hell to some, others embrace the possibility of aligning their bodies with productivity goals. Algorithmic involvements reproduce experiential digital divides that explain why the same optimizing practice can be both self-enhancing and self-depreciating, a theme that is further highlighted with analysis of the oppositional structure of feeling in the next chapter.

3 *The Digital Geography of Fear*

Maisa knows very little about algorithms but feels their presence in the form of intruders or even stalkers. She tells a common story about researching mobile phones on the website of an online store; as soon as she closes the website, ads for the same phones appear on her Facebook page, which irritates and scares her a bit because it feels as if big brother is surveilling her. When Maisa uses digital services, she does not read their privacy policies or terms and conditions of use; in any case, even if she did, she might not comprehend the actual and potential data practices they detail. Her feelings of apprehension are related to her insecurity about the nature of the information that is collected on her, by whom, and for what purpose. Companies' surveillance practices become experientially intertwined with the social risks of information sharing (Hargittai & Marwick, 2016). As far as Maisa is concerned, it is not only big brother that she needs to worry about, but other people as well. She fears that social media users whom she has not even met might be spying on her.

This chapter documents personal experiences of distress and insecurity connected with algorithms and related data practices. Articulations of fear and distress are treated as material for uncovering an oppositional structure of feeling, characteristic of algorithmic culture. Whereas the previous chapter focused on pleasurable engagements with algorithmic systems, highlighting the sense of enhanced autonomy and anticipation of improvements in personal life and in society, in this chapter the emphasis is on a reactionary stance. The confidence that people feel about technological futures shifts onto a terrain of uncertainty and distrust, underlining that we cannot be sure that algorithmic techniques are on our side. One person after another relates that they feel that they are no longer in a position to set personal boundaries; their narratives address the distress they feel when they cannot decide who can enter their private space, and on what terms.

Fear is typically felt when the corporate machinery either reveals that it "knows too much" or leaks information, unexpectedly destroying an illusion of privacy and security. What you thought of as intimate or "invisible" action becomes public, visible, and exposed. The interviewees talk about "wait a minute moments" (Bucher, 2017), as they infer algorithmic associations made based on their behavior; after reserving a hotel room, for example, an ad on a social media page immediately tells them of a festival in the same town. Mikael, a student of social anthropology, apprehends that the goal of such suggestions, enabled by algorithms, is to augment the flow of daily life, but they still feel "potentially dangerous." It is not so much that they are threatening in the here and now—the advertisements per se are fairly harmless—rather that they expose the predictive mechanisms in place, raising questions about future developments and the impossibility of knowing what lies ahead.

This discussion zooms in to emotional articulations that appear consistent and patterned, to query what they might tell us about the cultural shift in contemporary society that promotes affectively charged technology relations. I suggest that in relocating experiences of fear from the personal to the collective sphere, one route forward is by way of analogy; thus, I introduce the notion of a digital geography of fear to bring distress under a joint banner of structure of feeling. This opens a perspective on how people describe and cope with the fact that they have limited knowledge and control over the dissemination and use of personal information. Attending to the affective infrastructure invites us to see fear and distress in a new light, as a form of collective harm. Yet the way the respondents of our study discuss fear also makes it obvious that this oppositional structure of feeling is not evenly shared by all. As with the joys of technology engagements, the digital geography of fear accentuates experiential digital divides. The most confident articulations of digital distress comprise predictions of a dystopian future and present a linear way forward: warnings of how things will only deteriorate. Personally felt fear is typically discussed in a more cautious and restrained manner. Apprehension appears to be most intense among older interviewees and those with a more detached relationship to technologies, who fear that if they fail at some point in their technology use they will be exploited, taken advantage of, and harmed. The intimate and uncertain nature of feelings is stressed by their framing with precursors such as "I think" and "it feels like." The mundane and repetitive nature of algorithmic engagements can disguise how heavy the emotional burden can feel when machinic encounters touch upon intimate affairs and personally felt vulnerabilities.

Dimly Lit Parks of the Digital

At the end of the 1980s feminist scholar Gil Valentine (1989) opened a debate about the geography of fear that shadowed women's experiences in public spaces. She drew attention to how the fear that women feel in public spaces is a consequence of unequal gender relations and the associated lack of public safety. Women walk through dimly lit parks without knowing what lies ahead and fearing violence, because their gut feeling is that the rootedness of male dominance in society means that the possibility of attack is ever present. Statistically, however, in Finland as in many other places, violence against women is more likely to occur in the home, and the perpetrators are more likely to be men with whom the victims are familiar than random strangers (Koskela, 1997). The fear that women feel is thus not statistically accurate, but rather a reflection of a more encompassing societal structure that maintains gender inequality. All gendered violence, whether it takes place at home or in public spaces, and irrespective of the gender of the victim, can be traced back to social stratification and related inequalities.

What matters in terms of the gendered geography of fear is less the location of actual violence than the structure of feeling associated with it: the fear affects women and their notions of public spaces and society. It is not too far-fetched to point out that, similarly, in the digital world fear is a consequence of unequal power relations, as the informational asymmetries trigger experiences of vulnerability, exposing people to the possibility of being harmed. Before feminist scholarship outlined the gendered geography of fear, it was typically considered natural that women could not walk through a park alone at night, that when out of the house in the dark, unattended by a male, they were engaging in risky behavior; this is still the reality in most parts of the world, where young women in particular feel unprotected in public spaces. The notion of the geography of fear, however, helps to pinpoint fear as a collective experience. Relatedly, the digital geography of fear engages with personal experiences to provide a sense of collectively shared feelings that sustain the affective infrastructure of algorithmic culture. By tracing articulations of fear, distress, and insecurity, we begin to see how various kinds of algorithmic relations can make people feel vulnerable and exposed to possible harms, suggesting that the specter of fear and the related feelings it engenders are broad in scope. Thus, digital services can be thought of as the dimly lit parks of technological advance, activating feelings of potential violation. Articulations of fear are associated with data extraction, algorithmic operations, and larger processes of datafication, as internet users lack effective means to protect their private and public affairs (Draper & Turow, 2019). In light of the digital geography of fear, it does not matter exactly what companies do in a technical sense, or whether the reactions of users are based on accurate facts. The main point is that many people feel that service providers are blatantly violating notions of personal autonomy and privacy, which is what the digital geography of fear is signaling.

As suggested in chapter 1, the main methodological incentive for focusing on structures of feeling lies in the openness of the approach: feelings can be used to query and order disparate social forces. Fear aids in identifying aspects of algorithmic culture that are taken for granted, neglected, and not sufficiently explored. The digital geography of fear consists of experiences, ambiguous and ambivalent in their nature, ranging from the uncomfortable and scary sensation of being vulnerable and exposed that is triggered by violations of intimacy to the uncertainty or "mild paranoia" of not comprehending or having control over the personal or societal implications of datafication. When searching for evidence of the digital geography of fear, we cannot attain a comprehensive conviction of which experiences should be included under its banner. Feelings will always remain private, fleeting, and, to a certain degree, ambivalent. Combined, however, the articulations of fear testify to a structure of feeling that calls for public recognition of the opaque and privacy-violating nature of dataveillance and the feelings of fear and distrust that accompany it.

Becoming Resigned

Ella, a student of ecology, talks about a private conversation in which she and her friends were criticizing Neste—one of the biggest corporations in Finland in terms of revenue—which specializes in producing, refining, and marketing oil products. After discussing its ethically problematic practices, they moved on to the lighter topic of an online clothes cupboard that they were planning to set up. Soon after the conversation took place, Ella noticed an ad on her Facebook feed featuring clothes made by Neste. The incident was uncanny, especially as she had not known that the oil corporation manufactures clothes. Discussing the sequence of events with her friends, they shared similar experiences and weird associations, unexplainable unless mobile phones listen to private conversations. Ella says that the stories make her wonder how normalized the practice of listening has become.

Cecilia, who works for the National Broadcasting Company (YLE), says that she has not sufficiently researched whether it is true that microphones on cell phones record ambient audio to obtain private information to help in targeting advertisements, but she too has a story to tell. One day she was talking to her partner about a particular juice she enjoys and, when he misheard her, she repeated the brand name in a clear voice. Soon after, an ad for exactly the same brand of juice cropped up on her Facebook page. Our research participants repeatedly told us stories about such strange coincidences and eavesdropping practices, offering firsthand evidence of the concerns that people have in terms of data gathering, but also condemning the normalization of corporate practices of surveillance. The stories built a hierarchy of trust among social media sites with Facebook at the bottom of the pile; even if Mark Zuckerberg publicly denies eavesdropping, people reckon that it takes place anyway, because it is exactly the sort of thing that Facebook could do.

Whether Facebook or related commercial enterprises listen to mobile phones has been debated for years. Various unofficial tests and studies that have been conducted to find out, however, have concluded that phones are not an eavesdropping medium. Even if it were technically possible, researchers have not been able to validate such listening in controlled conditions. The possibility that specific ad keywords could be untangled from natural speech is also something that is repeatedly discussed, yet there is no certainty of which company does what. Experts underline that mobile phones do not need to listen to private conversations, because detailed personal information can be collected by other means, and with less hassle (Martinez, 2017). Facebook has adopted practices that have built on decades of experience with direct-mail consumer marketing (Turow, 2012). Carolin Gerlitz and Anne Helmond (2013, 1361) describe "the like economy" in their analysis, noting that it "draws attention to the Facebook platform and its back end data flows in which logging out, deleting one's profile or never joining the platform do not offer solutions to opt out." The Like button sustains an infrastructure that allows Facebook to collect data on everyone who visits any site with the button, whether they are registered Facebook users or not. This means that when people navigate internet sites, social buttons linking back to Facebook, such as the Like or the Share, send data about site visits back to the corporation. The collected transactional data, detailing people's online movements, can be attached to user profiles, instantly mined, and multiplied.

In practical terms, the constant shadowing of online traffic and phone movements is a much more effective way to scan what might be of interest to people than analyzing their private conversations. The way people talk is extremely complex, complicated in Finland by the Finnish language, which poses further challenges to any models or keywords lists that the company might be using. Personal stories about invasive data gathering, however, concentrate on eavesdropping via phones, rather than use of the Like button or analysis of location data. "Our phones are listening to us" has turned into what Brian Massumi (2010) describes as an "affective fact": it persists and even thrives on the debunking of the facts to which it is attached. Stories of strange coincidences that cannot be explained without the trope continue to flourish. The structure of feeling, sustaining fearful reactions and distrust, is strengthened with each new story, shared in person or by the media, signaling suspicious company practices.

Henrik, the life coach, relates that the risk that phones might be listening has shaped his behavior. If he is planning to do something that he does not want exposed, he puts his phone in another room or goes into the woods and leaves the phone behind. The potential for eavesdropping has aroused in him a defiant, child-ish desire to resist and mess up the Facebook algorithm. He shouts random things at his cell phone and uses these words as clues to test whether Facebook is listening. It could be anything, like yell-ing at the phone, "I sure feel like traveling to Lapland." If nothing happens, he might add, "I would love to see the northern lights." After repeating such sentences, he carefully reviews targeted ads to see if they start offering Lapland vacation packages.

In a perfect world, one could not even imagine one's phone eavesdropping on one's private life, as strict regulations would be in place governing how information is collected. Yet like many others, Henrik is almost fatalistic about the current situation, say-ing, "Society will never get the upper hand with this." His frustra-tion resonates with research findings that underline how powerless and resigned, apathetic even, internet users can feel in the face of digital developments (Hargittai & Marwick, 2016; Draper & Turow, 2019). Henrik concludes that even with the GDPR being enforced in the European Union, things are not likely to improve. In social media, the situation continues to be messy, because informed con-sent forms the basis for data processing. The consent should be an unambiguous affirmation by the data subject, but in practice con-sent might be given by default because people want to use the spe-cific service as swiftly as possible.

The notice-and-consent model fails to ensure fair and trust-worthy practices, because it does not build on the manner in which people actually engage with digital services (Draper, 2017). Deci-sions about cookies, or associations with data practices, are made with inadequate information, and opting out is not considered an actual choice. If Ella, for instance, had knowingly consented to current data practices and could foresee all possible consequences, she would have no need to share stories of intrusive media and privacy infringements with her friends. In the current situation, however, companies do not have to do anything illegal; they can simply take advantage of the fact that people click their consent without reading the terms and conditions of services. The data collection practices are legal, but data practices can still feel scary and creepy. The structure of feeling is an aftermath of lawful practices, which feel experientially illegitimate.

Anxieties after Consent

The notion of consent, granted separately to each website and service, obscures the fact that it is almost impossible to comprehend all that one is agreeing to. This means that people continue to have doubts and anxieties about data practices after they have given their consent (Tanninen et al., 2022b). It might not be the data extraction or the loss of privacy that disturbs people, but a much wider unease. Like women who feel nervous in dimly lit parks, people online might be haunted by the impossibility of feeling relaxed and safe in the digital world. They fail to find the affective alignment, temporarily or more long term, that characterizes pleasurable engagements with algorithmic systems.

When we raised the issue of the insecurity that people feel in relation to algorithmic systems in the interviews, technology and marketing professionals treated it as a consequence of technical illiteracy. If people knew more about algorithms and had greater data literacy, they would realize that nobody is stalking them or listening to their phones. The machine is merely linking data points, and at times it succeeds in connecting the dots so accurately that the result feels intimidating. Simon Pitt (2020), the head of corporate digital at BBC, treats "wait a minute moments" that people experience as evidence of the quality of humans as "pattern-spotting machines." It is not only machines that connect the dots; people connect them too and might detect patterns where none exist, filling in gaps in their understanding with narratives that help them make sense of what algorithms do. They might, for instance, see harmful causalities when algorithmic recommendations are too accurate and expose their vulnerabilities.

In a *Wired* article, Antonio García Martínez (2017), a former product manager of Facebook's advertising targeting, describes the deluded human pattern of equating "what we would most hate to have revealed with what advertisers (or Facebook) would most like to know." Arguing from the position of digital marketers discussed in the previous chapter, he states that marketers do not handle user data as personal. Consumers are not treated as bounded entities, positioned in predefined consumer segments, but as dividuals, who as Gilles Deleuze (1992, p. 4) points out, "continuously change from one moment to the other." Someone's data—whether concerning the location of their activities, expressed interests, or purchases—is dynamically organized, defined, and assigned value by means of automated processes that aim to influence behavior in ways that benefit the market. Consequently, Martínez describes the assumption that our personal lives are important or interesting to companies as a "narcissistic fallacy." Only datafied parts of those lives, features that can be digitally combined and manipulated, are useful to marketers, details that might have nothing to do with our vulnerabilities, the deepest insecurities and secrets that we would like to keep to ourselves.

As will become obvious in the next chapter, people tend to approach questions of targeted advertising in an ego-centered manner, from the position of their own intimate and private sphere. They evaluate experiences with ads based on how they feel about them. Yet the notion of the digital geography of fear suggests that instead of labeling fear and discomfort a narcissistic fallacy, they should be acknowledged as forms of collective violation. What Martínez and his marketer colleagues fail to understand is that even if personal data does not appear personal to the digital marketer, people whose lives are being datafied live closely with the technologies. They coevolve with technologies and become intimately tied to data collection procedures: data that might disclose their sleep, health concerns, menstrual cycles, or quest for sexual partners. Data is not personal in the sense that private property is personal, despite being discussed as such at times, but people can still feel ownership of their data (Maurer, 2015). Data traces belong to the person who leaves them, because they are traces of what that person believes in, desires, or has done in the world. Furthermore, the digital marketers' dismissal of the personal nature of data should not obscure the fact that digital marketers do care about personally significant data—way too much. Digital marketers are all over people as they collect browsing histories, shopping behaviors, and location data with the purpose of targeting individuals with their messages. Even if the data extraction is done en masse, in a nonindividualistic manner, it can still hurt people.

Obscurity as a Market Logic

Once the fear and paranoia have started to crystallize, contradictory evidence will no longer change the state of affairs. The fears related to data extraction and algorithms might initially be bred by lack of accurate information, with people forming their opinions based on skewed or missing facts, yet the distrust also derives from the ensnaring nature of digital services; as people start to use such services, they voluntarily, if unwittingly, become enmeshed in the webs of information that fuel service operations. For example, around 2015 Finns began ordering 23andMe genetic test kits, which gave them access to personalized ancestry maps and charts listing elevated health risks. This was done with little understanding of the less public agenda of the company, which was to get consumers involved in the production of a database that can be used for building corporate partnerships (Ruckenstein, 2017). The way the market logic of direct-to-consumer genetic testing is hidden, including the commercialization of health data and participatory research initiatives, speaks of a more general characteristic of companies, that of concealing their value extraction mechanisms. The rhetoric of openness and participation widely promoted by digital platforms obscures the lack of transparency regarding economic pursuits related to uses of personal information.

Ultimately, it is difficult to acquire firsthand information about value extraction mechanisms other than from company-initiated blogs or news sites. When people recognize the extractive uses of their data and how their participation is integrated into the business logic, which sometimes happens long after they have started to use a service, they can feel betrayed. For example, once people realize that the genetic data uploaded to the 23andMe service can be sold to pharmaceutical companies, with no financial dividend for the test takers whose data traces make up the database, they start to look more carefully at how the service actually operates. Ironically, with their test purchases they had paid a fee to become contributors to resalable health data sets. In light of the oppositional structure of feeling, it is thus notable that breakdowns in trust direct attention toward "unpicking how the system works" (Bishop, 2019, p. 2592). Only after the trust is lost do people begin to pay attention to what actually happens with the aid of data they have supplied.

At times, feelings of being deceived are directly related to the privacy settings of services. Sara recounts how she participated in a group discussion on Telegram, an instant messaging app that is advertised as a more privacy-friendly service than WhatsApp. After exchanging numerous messages on the assumption that they were private, the group of friends suddenly realized that group Telegram discussions are searchable by anybody. It turned out that in order to have a private conversation, one needs to switch on the appropriate settings. Sara describes feeling shaken, because she and her friends had written openly about their private lives. What this example brings to the fore is that privacy is not a default setting, but something that people need to guard and protect on their own initiative. The service, even if privacy protecting, operates on the initial assumption that people in groups need no privacy. As it turned out, the administrator of Sara's group had in fact secured privacy for their discussion but, with trust lost, the group decided to move to Signal, yet another messaging service.

The muteness of platforms underlines their position of power: users are subordinates who have to accept their default settings and decisions. Unless you are exceptionally well positioned—for instance, a celebrated influencer—communicating with digital platforms such as Facebook, Instagram, or YouTube is like talking to a wall. Every day people all over the world send queries and requests to these companies, trying to understand why the platforms and their algorithmic functions operate as they do—another kind of collective experience of our times. The silence of digital platforms nurtures a culture of not-knowing, a perfect breeding ground for algorithmic folklore, stretching from rumors and gossip all the way to elaborate conspiracy theories, as people try to alleviate the uncertainty and powerlessness they feel. Frank, the growth hacker, perceives information sharing with peers to be a slippery slope. It might help people cope with the platforms on a practical level, but by relying on speculations about how algorithmic operations work, they might end up basing their observations on conspiracy theories and ill-advised associations. In the end, this might only nurture the insecurity and distrust they feel.

Ella thinks that her fears have at least partly to do with her inability to understand why digital services, including YouTube and Facebook, have such a big influence on her, hooking her in; indeed, to her they appear to possess controlling powers. Yet because of the opacity of digital services, algorithms do not readily reveal their powers. Ella fosters dystopian thoughts that Facebook is used by states for governing their citizens. How many Cambridge Analytica scandals are presently in the making? As we only learn of violations after the fact, we have no way of knowing if they are already on the way. Leo, with experience in a cybersecurity company, continues this line of thinking. Since we cannot tell how algorithmic processes are currently used—and only find out about them in retrospect—he suggests that it is possible that a political party, or some other governing entity, is manipulating people's behavior as we speak. These articulations of concern and others like them suggest that companies and organizations have the ability to use their resources for increasing the scale of manipulation. As knowledge is power, algorithmic governance turns into a potent weapon if consistently used for manipulating and controlling others. Whether the companies decide to support that aim, we will typically only find out in hindsight, if at all.

Personalized Safety Zones

As demonstrated in the previous chapter, the joy found in technology use is strengthened in proportion to the sense of autonomy and exploratory attitudes that people have. In contrast, fear sweeps away self-determination, making people doubt themselves and their ability to cope. In order to strengthen their autonomy, the respondents in our study report privacy-related practices they adopt to armor themselves against possible harm. Ad blockers are used to avoid targeted advertising, and anonymity is protected with virtual private networks (VPNs) and noncommercial services like DuckDuckGo, a search engine that, unlike Google, stores no personal information. Some users hide their IP addresses, alternate between different browsers to shield personal data traces, reject the use of cookies, use multiple social media accounts and profiles, attach the accounts to different email addresses and phone numbers, or use fake names. These acts are aimed at obfuscating and rejecting companies' data harvesting efforts. It might be impossible to know how effective these practices are, but on a personal level they are important in terms of reinforcing the feeling of being in control. The withholding and curating of personal information to enhance privacy and evade companies' profiling attempts offers a means to protect, sustain, and reclaim personal autonomy. As they gain experience, people can develop their own "privacy infringement maps" that help them to navigate the digital world and identify high-risk places and practices to be avoided.

Privacy guarding practices, and services that support them, are a direct response to the digital world that leaves questions of privacy in the domain of individual action. This has naturally produced a market for tools that will take on the task on behalf of the individual, notably VPNs, which make online actions virtually untraceable. Max, who defines himself as an active user of digital services, tells us that he will purchase the Finnish cybersecurity company's Freedome VPN as soon as he can afford it. The service promises safety from "hackers, trackers, and intrusive companies." It hides users' internet protocol (IP) addresses and online traffic and offers detailed information on who is tracking them online. Privacy services and individual acts of obfuscation strengthen the boundaries of privacy—*my* privacy. Services that protect privacy as well as concepts such as informed consent isolate people from each other and treat them as atoms that need to create and nurture their own safety zones, positing that every individual is responsible for taking care of their own protection. Like the women who drive rather than walk in the evenings to avoid dimly lit parks, users of these services find work-arounds to allow them to feel in control on their own account, much as, in the safety of their cars, doors firmly locked, women can disregard those who need to walk home from the metro station via dark alleys and underpasses after a long day at work. However, just as cars as physical safety zones protect the individual but not the vulnerable group, individual privacy-enhancing acts that address cyber-threat susceptibility fail to challenge the data power at play. When people combat the digital geography of fear with technological savvy or with services that promise to protect privacy, the informational asymmetries are left intact. Outside of personal safety zones, people continue to feel afraid and insecure, and tell stories of how their phones listen to them.

A side effect of privacy-preserving services is that they promote feelings of further inadequacy. Ultimately, the risks are delegated to individuals, who are advised to make rational choices concerning matters of personal safety. Many interviewees regret-ted that they were not more proactive, acknowledged that they had done too little to protect their privacy, and wished that privacy-protecting services were more affordable and easy to use. People might use a network such as Tor, offering nonprofit browser pro-tection, in order to surf the internet anonymously, if they are not comfortable with the global machinery of data extraction of which social media services like Facebook are part. Mikael, however, echoes the opinions of others when he talks about the impossibil-ity of becoming a regular user of Tor, noting that it is cumbersome to use and would complicate daily life. Tor, Freenet, and peer-to-peer networks operated by public organizations and individuals function in the "dark web," where people can withhold their iden-tity and communicate and engage in business transactions with-out exposing their location. The dark web forms only a fraction of the deep web, which is not indexed by search engines. In everyday talk, however, the dark web is often conflated with the deep web and, despite their differences, both are associated with drug and firearm sales, gambling, and child pornography. Consequently, the idea that in order to use the internet safely one must follow the routes of criminals and pedophiles does not appear reasonable. It is as if it is not "normal" to protect one's privacy, if one needs to join deviants and transgressors to be able to do that.

Choose Not to Fear

The digital geography of fear comprises assessments and valua-tions of what is happening, as well as what is possible for whom and under what conditions. The discomfort with current data-gathering practices shapes the experiences of those who feel that they cannot master digital tools and have insufficient skills to boost their autonomy in the digital world. Yet fear and distrust also separate experientially optimistic technology professionals from professionals with a more suspicious or cynical attitude toward digital technologies. In order to maintain a positive vibe, the participants in our study might knowingly distance themselves from the potential risks and harms of future developments, instead stressing their agentic possibilities. Rather than recognizing the crippling effects that the digital geography of fear can have, they emphasize the need to shake off tensions and act. As professionals sustain and actively perform the pleasures of technology development, they deal with the insecurities associated with dataveillance by mitigating their effects. Henrik is a case in point. As we learned in the previous chapter, Henrik nurtures a positive online atmosphere by cleaning his Facebook feed of negative influences and promoting optimistic messages about technological advances. While he acknowledges the threats to privacy and democracy that digital technologies sustain, he refuses to get stuck with fears or even pay much attention to them. He says that he cannot take "a victim attitude" and start bemoaning how horrible it is that his personal data is being exploited. From his perspective, this is a logical approach: he could not feel as excited about digital developments if he focused on anticipations of the negative. For him, the division of labor is clear: while he concentrates on the constructive future scenarios appreciated by the business world, others can deal with the mirthless sphere of risk and harm.

Yet when he stops to think about the surveilling powers enabled by extractive data-gathering practices, Henrik starts to feel distressed. Both professionals and nonexperts oscillate in the interviews between negative and positive future scenarios, depending on how much agency and optimism they feel they have in steering future developments. Thus, they are simultaneously living different versions of the algorithmic future. If, for example, Henrik wanted to practice some form of civil disobedience, he would have to take into account that the opponent—"be it a community or a state or whoever"—can track his whereabouts. At the very least, the authority that maintains his local public transportation network would be able to identify his daily routes, based on his travel card data. He observes, however, that in Finland the possibility of the state or other official entity using his personal information frightens him less than system failures do. He refers to a Facebook-related story that broke in 2012 that a bug within the site had started publishing archived private messages on users' walls. The story is probably not even true, but it continues to live as algorithmic folklore. Henrik says that the possibility that private conversations on Facebook could become publicly visible scares him, because he has had online sex. The thought of his intimate exchanges becoming exposed horrifies him. During sex, he could have switched from Facebook chat to another channel but eventually chose to remain in the chat because, as he says, "I was already there, so be it." His experience testifies to the difficulty of fully controlling one's own actions or the digital space as lives move more permanently online.

Heidi, a business school student, shares the opinion that not all control mechanisms are desirable: people should have the freedom to act and do things in ways that others do not know about. Her thinking emphasizes, however, that personal autonomy is not a fixed state of affairs but should be actively pursued. People need to become more proactive and learn to dissipate anxieties, as the only way forward is to learn to live with the insecurity. As she points out, "You can, of course, just wipe out everything—at least your own behavior and your own profiles on social media and stuff like that—and move to live in a hut in the woods, with no electronics around. Is that very smart? No." For Heidi, remaining stuck in one's fears equals becoming marginalized: one is no longer an active part of future developments but remains frozen in time, unable to participate in the digital society. Whereas the coevolving with algorithmic systems creates an undercurrent that moves things forward, fears slow them down, creating barriers and hindrances. Ultimately, the experiential landscape of the digital geography of fear is contextual and situational, and forward and backward movements are evaluated and reevaluated.

Marginalizing Insecurity

In the interviews, the digital geography of fear was sometimes made tangible when participants listed actual and potential harm and distress. This extends the structure of feeling fed by insecurity and paranoia to a broader array of possible harm that algorithms and technologies can cause. Liisa, a practical nurse and a nutritionist, relates that first her family's bank card was stolen, then her husband's credit card was hijacked and somebody made purchases with it abroad. She also remembers that a few weeks earlier the printer had started working in the middle of the night, producing bizarre copies; somebody had hijacked that too. In the end, it might not be the data gathering itself that worries people, but rather damaging data movements and uses, particularly those that are not company initiated (Lupton & Michael, 2017).

Erika, who works as a chef, says that she is "mildly paranoid" about algorithms and the uses to which private information is put. She connects her paranoia to poor mathematical skills and a lack of understanding of how machines function as she describes the nagging feeling that follows her: "If I've visited a page once to check some things and I don't necessarily want to buy them, they will stay and haunt me. 'Njah, you were here.' It is as if there is a mommy-like being, rubbing it in, who is like, 'Oh, you went to see these and these.'" After Erika describes the "mommy-like being" watching and stalking her, she turns to other fears she has, including frauds, scams, and identity theft, adding that she contains her uncertainties by engaging in privacy-preserving practices. Unlike more tech-savvy users, however, who have detailed knowledge of how the services work, her practices are based on what she assumes is needed for protection: she joined Facebook with a fake last name, is watchful of the information she shares, and obfuscates potential "stalkers" by not revealing her geolocation. Erika says that one of her friends has even greater fears about data protection and identity thefts. He does not post anything on Facebook that might expose his identity. When he meets new people face to face, he does not even tell them his real name. Erika is happy that she is not "as paranoid," yet as she continues her story, it becomes apparent that her anxieties are also difficult to suppress. For her, fear is an experience that frequently captures her attention, affecting her overall relationship to digital technologies. She is very wary about burglaries, which she feels is possible if she updates on Facebook that she is on vacation. She is worried about technology addiction and relates how digital technologies affect the quality of her sleep, spoil her concentration, and make her "stray onto stray paths." She suffers from fears of radiation, caused by Wi-Fi in particular, and claims she can feel the radiation when she holds her cell phone. It gives off a "wobble" and, especially when she charges it, she can feel a "kind of vum-vum-vum" as the energy flows in.

Digital insecurity might be seen as shameful, talked about somewhat apologetically. The digital geography of fear continues to be culturally associated with those who are unable to reap the benefits of digital developments. "Back to the stone ages," Mikael says when he refers to a societal need for an anti-algorithm lobby, while those who fear and distrust digital services sometimes describe themselves as "ostracized" or "pensioners," indicating their peripheral status in digital society. One young woman calls herself a granny, another one a retiree, illustrating her marginality with her refusal to download Mobile Pay. She mistrusts the service, as it would require her to save bank details on the phone, a view she sees as a sign of not being quite up-to-date and, hence, a retiree. Digital distrust is seen as somehow reprehensible even among those who feel it, because it suggests that they do not live up to societal expectations in Finland, one of the most digitalized societies in the world. Ella feels that, despite her youth, she is turning into a "backward grandmother fighting against development." Instead, she should trust that if algorithms truly were as awful as she thinks they are, people would flock to the barricades, adding, however, that her idea is "a bit stupid" because there is so much in the world that people fail to resist. Ella admits that she is stuck with the idea that digital technology is bad, but even if this positions her as culturally backward, not really belonging to the digital society, she cannot shake it.

A common way to refer to the societal mismatch is to talk about "being a foil hat" or moving to the "foil side," signaling that one's thinking might be paranoid or dystopian in light of actual developments, that one stands out societally and has an inadequate relationship to technologies. As Erika concludes, "How much is really true and how much is paranoia? Where can the line of the famous foil hat be drawn? Not with me, not quite. Personally, I would not put on the foil hat." Pauli, who is generally trusting in his approach to digital technologies, cautions that he might be going to the "foil side" when he considers how algorithms are used to influence how people think. Rather than localizing digital insecurity in individual psyches, however, he emphasizes the larger societal processes at stake, underlining the need for a broad notion of the digital geography of fear. For Pauli, digital insecurity is connected to states and organizations manipulating and instilling insecurity in others. Data companies can hand over information to US intelligence agencies, while intelligence operations are used to advance state interests in fields as diverse as trade policies and security concerns. Russian hackers build fake news sites and use algorithms to bombard people with misinformation. Pauli emphasizes that propaganda is not a new phenomenon, but in digital spaces it might be harder to figure out what is propaganda and what it is trying to do, which shakes the foundations of democratic societies, spreading webs of anxiety and insecurity in multiplying forms. Growing anxieties are experientially connected to local ways of expressing distress and fear, a theme that I turn to next.

Heating Up

Fear is not necessarily what younger Finns talk about when they describe the distressing aspects of datafication and the digital world; instead, they use the verb *kuumottaa*—heating up—a word that has no direct translation in English, as it is a local expression that refers to an emotional state or affect. Originally a subcultural slang term, heating up became mainstreamed with a popular Finnish travel documentary called Madventures, which premiered in

2002. In their travels, the Finnish TV host and his compatriot cinematographer used the expression to describe how they felt about obscure local customs. Today, heating up is associated with a wide range of situations, events, and potential future developments. In the Finnish everyday, heating up can describe being tense about a job interview or fearing the consequences of one's actions. For instance, if you go into a grocery store but they do not have the item you are looking for, you feel a heating up as you leave the store empty-handed in case you are mistaken for a shoplifter.

In the context of the digital geography of fear, heating up is related to various kinds of harm and vulnerability caused by digital developments and the broader societal implications of datafication. It indicates an awareness of the negative effects of processes of datafication and the frustration that little can be done to mitigate and improve the current situation. The flexibility in application of the term makes heating up a powerful concept with which to express the local character of algorithmic culture. It can also be used humorously, or in an ambivalent manner to avoid sounding as if one is sliding toward "the foil side." Heating up can also be felt on account of the fear that others might be vulnerable to digital harm. Kari, a biology teacher who is careful with his own data practices, says that he feels a heating up because of his mother's habit of clicking on random links and being clueless with her personal information. When offspring worry that their aging parents might harm themselves, they often take over digital tasks to shelter them; alternatively, the parents might actively expect this. Liisa, in her fifties, says that she has never become cognizant with digital technology because her son became her information and communication technology support as a child. From the age of ten, he has been researching things for her online and conducting digital tasks.

Children who heat up on their parents' behalf know that little empathy is directed to elderly people who click fraudulent links because of limited computer skills or poor eyesight. They dread the day their parents upload malware onto their computers or accidentally share their bank information. The need to protect their loved ones connects to the way individual behavior is questioned in cases involving violation of privacy. As with the gendered geography of fear, victims are often blamed for their actions. Teresa, in her late thirties and working in the sales department of a large technology company, considers users the biggest problem in terms of fraud. She speaks from the position of a technology professional for whom digital harm is caused by erroneous and ill-informed humans who circulate sham links when the local supermarket promises to hand out a €500 gift card. "Or they think that a Nigerian prince is waiting for them. Well no!"

Heating up connects to personal actions or the behavior of others, but according to our material, most typically it is associated with technology-related developments and the impossibility of knowing what lies ahead. Heating up is the feel of an uncertain future. It is a bodily response to the existential threat of the unknown. Nina, currently unemployed, cannot pinpoint any concrete misuses of personal data, but she attempts to articulate an overall negative outlook with the notion of heating up. She ends the interview by asking what the right reasons to fear the collection of data would be. The oppositional structure of feeling, linking articulations of fear with experiences of heating up, is fed by thoughts of dread about the effects of data gathering, even if they remain unidentified.

For Frank, the growth hacker, heating up refers to paranoia, yet perhaps to some extent legitimate paranoia. He positions himself in the role of an adult who observes others' heating up in a detached manner. He explains it by drawing an analogy with the tendency of the terrible twos to say "no" to everything. This negating inclination, although childish, is active in all of us; we react if our personal autonomy is endangered. Referring to George Orwell's political dystopia in *1984*, Frank says that unlike in the book, contemporary manipulation is not big brother-type surveillance that aims to hijack and control people's pasts and presents; rather, it is aimed at directing action. The main goal is to speed up the process of making purchase decisions. Others, like Sebastian, however, underline the political-economic ramifications. In contrast to the joys and pleasures of technology development, which promote a sense of autonomy and anticipation, heating up connects to an inability to mitigate structural forms of harm. Sebastian, a law student, locates heating up in the context of the excessive data gathering enabled by mobile phones. He describes how "you get a bit of sweat on your forehead when you watch documentaries or whatever on the subject; yes, it starts to heat you up." He talks about the difficulty of comprehending how a little device, a mobile phone, can know so much about you. "I am somewhat aware, but it is still difficult to discern it, the vastness of it all."

When thinking about the powerlessness to resist current developments, Sebastian connects algorithmic powers to a new kind of passivity that is much broader in its implications than the question of privacy: people are turned into "unreceptive parsnips that update the YouTube channel." They are shaped with the techniques of persuasive design, hooked and trapped with recommender systems. He says that as long as young people are kept entertained, they are comfortable not knowing or understanding the bigger picture. The global "devil companies" have generated "a ready-made package," a world that no longer has anything to do with innovation for future generations, but rather tries to keep us passive and submissive like vegetables, "digitally resigned" (Draper & Turow, 2019).

Sebastian talks about a new world order that values technical authority and expertise above all. In this world, he is marginalized and lags behind because of limited Excel skills. His angry ranting turns him away from an obvious contradiction: as a law student at the University of Helsinki he is hardly the underprivileged individual that he depicts himself to be. Yet the way he positions himself in relation to processes of datafication harbors an emotional intensity that gestures toward a future that he feels is imminent. Capitalist developments coupled with algorithmic powers are moving us away from the social progress that could nurture and develop democratic societies. He sees bad choices and lost opportunities ahead, a view that resonates with the conclusions of critical data studies. As John Cheney-Lippold (2017, p. 257) puts it, "Our worlds are decided for us, our present and futures dictated from behind the computation curtain."

The People behind the Algorithms

Articulations of heating up relate to negative and dystopian views and worst-case scenarios in technology, but Cecilia thinks that in effect people are afraid of the professionals behind the algorithms. As she says, "It is such a faceless place where that information goes. Who is there, who uses the data?" Emphasizing the amazing potential of automation and robotics to raise societies to completely new levels, she points out that if people were truly freed of mechanical work, as the techno-optimistic scenarios promise, there would be room for creative thinking about the future, which is urgently needed in current climate crises. Yet we cannot even get close to such potential, because the power of designing technologies and thinking about their applications is restricted to a shrinking group of professionals: a few privileged white men. For Cecilia, heating up consists of generalized anxiety about the future going wrong in the hands of data corporations whose directors appear to have little interest in the fate of young people in globally peripheral places like Finland. States and citizens are inadvertently relinquishing sovereignty and power to corporate giants with enormous data resources as the professionals hired by these companies from top universities globally are using their skills and capabilities to steer our actions. While we are not fully aware of that loss as it takes place, our emotions and bodies suspect it, as they keep heating up. One way to think about heating up, then, is to treat it as an emotional resonance with our times, an internalized zeitgeist that embodies the critique of datafication. Heating up concretizes, on a personal level, worries and fears about public culture.

Nina, who mainly uses digital technologies for entertainment purposes, wants to believe that algorithms are useful, but she finds it disturbing that they constrain her worldview; she is worried that she might no longer encounter contrasting viewpoints on political and moral issues. Here, heating up is felt in relation to "filter bubbles" (Pariser, 2011), our personalized informational universes from which tensions and contradictions have been removed by the work of algorithms. Empirical studies have been conducted to both confirm and debunk the existence of filter bubbles (see Bruns, 2019), but in terms of how Finns talk about their algorithmic engagements, the experience of being in a bubble, and being shaped by a bubble, is very much alive. Nina says that her personal informational ecosystem is definitely more pleasant than some "racist bubble," but if that is all the information she ever encounters, then racism is so purified from her digital sphere that she might think that it is not an actual problem in Finnish society. Eemil, a microbiologist, literate in algorithmic matters, continues Nina's line of thinking and says that if the received information merely confirms one's worldview, it creates distance from those who think differently. Based on how social media algorithms treat him, he has been profiled politically as green and left wing. If that is all the information he gets, Eemil asks, how can he understand other political ideologies? A man who only converses with his mirror image quickly turns into a one-dimensional man.

Received information has of course always been confined to a certain extent, but as the respondents of our research repeatedly point out, the current algorithmic mechanisms have the power to accelerate informational bubbles and echo chambers that feed into what one already knows and has expressed interest in. Ella comments that instead of promoting a free flow of information, algorithms "offer it ready-made." Algorithmic processing of information can cut of slices of history, in a manner that flattens the past. This kind of packaging of information not only compartmentalizes it but can hinder contextual understanding and change. Ella wants to learn about new things and perspectives, and feels she is changing as a person, yet algorithms have a certain conservatism built into their logic and will continue to offer information based on what she has favorably acknowledged in the past. As she observes, however, "What if an invention that could revolutionize the world goes unnoticed, because it does not attract enough likes?" Because the tyranny of the algorithm favors what has already become popular and appreciated, the machine might not recognize the truly exceptional.

Shared Dystopias

The discomforts articulated with the notion of heating up resonate with the fear of losing common ground in how we appreciate and experience the world; algorithmic developments are seen to lead to an inability to learn from each other, as everyone builds a subjective impression of reality untouched by the experiences of others. Yet dystopian future scenarios offer an alternative common ground, one with powerful unifying narratives and symbols, such as the devious Facebook. Many of our interlocutors mention Donald Trump, the former president of the United States, in close association with algorithm-assisted social polarization. The way he was elected and how he used and misused social media in his governance exemplify active efforts to undermine democratic processes. Historically, Trump's trajectory is not an entirely novel populist phenomenon, but his brand is "nastier, crasser, and more belligerent" (Rosenfeld, 2019, p. 127) than earlier varieties.

Rosa, a human resources assistant, compares Trump's mode of influencing to that of junk news media, whose circulation benefits from emotionally charged content (Savolainen et al., 2020). In this light, algorithms are viewed as obstacles to healthy public culture, as they promote affectively engaging "black and white" actions. Like many others, Rosa regards Trump as a manifestation of today's polarizing developments, breeding misinformation, exaggeration, and outright lies as he actively promotes the gap between his supporters and opponents by communicating a destructive "us and them" divide. With their sorting mechanisms, algorithms further deepen the divides, accelerating conflicts and tearing people apart. To combat current tensions and developments, Rosa advocates algorithm development that favors scientific facts rather than affective intensification. For her, public culture is not threatened by algorithmic operations per se, but by the way the IF-THEN rules of algorithms are defined and employed.

While Trump has become a global symbol of algorithmic dissemination of aggression and misinformation, ironically, he has also become the archetypal character in a shared dystopia that defines Finnish algorithmic culture. Teresa connects the mechanisms of algorithmic dissemination of misinformation to Trump, but also to Russian troll armies—another threat to democracy—claiming that Finns fail to recognize an ongoing information war with their neighboring country. After Russia's invasion of Ukraine in February 2022, the perception of information warfare changed very quickly in Finland. Teresa, however, had already talked about it when few people took it seriously. Traditionally, such a war would be waged by means of propaganda, by the Russians dropping information leaflets along their border and announcing through loudspeakers, "Come, you will get bread." The current information war is "much more effective and smooth." Teresa doubts that Finns notice which comments are from Russian trolls in the online comment section of a Finnish tabloid newspaper. Her worst fear is that data and algorithms are being used to disseminate silent propaganda to mobilize further racist movements and raise support for the Nazis, adding that the crusade against women is already active in Europe. This is really scary, she thinks, because it could lead to revolutions: "We might soon be in Gilead," she says, making a reference to *The Handmaid's Tale* by Margaret Atwood. "This is the dystopia," she concludes, saying with a laugh that while, luckily, the troll armies are not smart enough to accomplish all this, it is important to maintain a healthy dose of mistrust and not to be naïve.

Shared dystopias are not distressing in the same way as personal experiences. They localize harmful developments elsewhere, distracting Finns from what could be shared real-life concerns locally, but also globally. For instance, a theme that gets hardly any attention from the respondents in our study are the environmental consequences of datafication. Eeva, a crafts teacher currently at home with small children, is one of the few who connects technological developments to ecological concerns. She regards the impossibility of maintaining continuous technological and economic growth as the most pressing issue of our time. For her, the digital geography of fear remains incomplete if it does not take into account that the technological infrastructures need natural resources in their production and use, and that they test ecological limits (Hogan, 2015). Eeva has no longing for the internet-free time of the past, but the current, ubiquitous digital world is not what she wants either. She thinks, or at least hopes, that our relationship with technologies will become more reasonable at some point; otherwise we will not be able to live well in the future, or at all. Thus, she transcends dystopia to present what is a logical conclusion to our continued expansion of the world of technologies without careful consideration of the ecological underpinnings.

Reclaiming the Collective

This chapter has offered the notion of the geography of fear as an analogical lens through which to explore a realm of affective infrastructure that is born of everyday experiences of fear, distress, and insecurity. The goal has been to demonstrate how collective experiences of fear and paranoia draw attention to present-day vulnerabilities. The notion of a digital geography of fear highlights the shared nature of insecurity and distress, as it speaks to the failure to create safe public space for all. An important difference between the two geographies, gendered and digital, is the nature of the threat that causes fear and distress. In the case of women and public spaces, the perceived danger is physical male violence, which by its nature is clear, direct, and concrete. The coping strategies are avoidance and alertness. Women scan their surroundings to detect risks and choose paths that have lights or other people; they might run past intimidating places.

In the case of processes of datafication, perceived threats are much more difficult to pin down. They are indirect, amorphous, and at times related to imagined rather than prospective future developments. Futures are felt and anticipated rather than known. Violations of personal boundaries repeatedly take place, but the nature of violation is not straightforward and requires an assessment of the possible loss and associated harms and risks. Such evaluation is complicated by the way data companies downplay their social and societal powers. Whatever power over the user a particular digital service has, it is not designed to be rough and explicit like physical violence; the rhetoric of participation and social sharing and the promotion of usability and frictionless digital living typical of digital services diverts attention from power inequalities. The manipulation exercised by corporate powers is meant to be subtle, gently guiding and nudging users to behave in commercially beneficial ways.

The lens offered by the digital geography of fear makes the patterned nature of personal experiences concrete as it focuses on the collective nature of insecurity and its links to underlining structures of power. Local emotional vocabularies are of key importance because they contain intimate knowledge of experiences with algorithms, enhancing analysis of how algorithms are felt and human-technology relations responded to. The articulations of emotional reactions, including the experience of heating up, push us to see vulnerabilities that are typically not attended to. One can think of the experiences of fear and distress described in this chapter as consequences of system failures, in the sense that they reveal informational asymmetries and related practices of power. Emotional responses, reiterated in personal reflections time and again, call for seeing beyond the individual, giving form to the felt effects of algorithms and ultimately raising more general questions about algorithmic agencies and how they should be dealt with collectively.

4 *Friction in Algorithmic Relations*

Kim and his friend Jon are using Jon's girlfriend's computer when suddenly the screen is flooded with baby-related ads. Their first thought is, "Oh my God, is she pregnant?" Finding this a worrying prospect, they speculate about possible online searches that might explain the targeting. In a related conversation, however, Heidi refutes the fears of pregnancy. She explains how annoying it is that baby-related ads rarely reflect actual life circumstances. She gets pregnancy test promotions because she is a young woman, not because she would like to procreate. As described in previous chapters, digital marketers treat users as data traces and mutable dividuals. The goal is to dynamically organize and assign value to the extracted information in order to identify consumption patterns in a relevant and timely manner. Yet targeting in digital advertising can still follow very basic segmentation models, relying on factors like age, gender, and location. Iida, complaining about gendered advertising, knows this and explains that being shown ovulation and pregnancy test ads based on her demographic profile irritates her, because she would like to "continue to live a wild youth, but society is deciding that it is time to settle down." She would be equally annoyed by ads in women's magazines proposing stereotypical gender roles, but they do not similarly invade her intimate sphere. While Facebook is an "advertising oligopoly" with promises of "accelerated time between product advert and sale" (Skeggs & Yuill, 2016, p. 381), it is still considered personal and social space, which explains why people react to targeted advertisements in the way they do. Advertisements enter their space, pushing messages that they have not asked for.

This chapter departs from these highly recognizable experiences in the quest to locate an emerging structure of feeling that addresses the tensions and ambivalence involved in algorithmic relations. Irritation is born when algorithmic systems are seen to fail to serve the aims of individuals as expected. The anticipation of future rewards from algorithmic systems turns into frustration when the promises of efficiency and smooth coevolution with machines remain unfulfilled. Not surprisingly, interviewees who understand the basic mechanisms of how algorithms work offer the most technically engaged and illuminating accounts of irritation; indeed, as discussed in chapter 2, such know-how increases people's expectations. The more closely people interact with algorithmic systems, the more they would like them to be developed further, a desire for improved functions that is an integral part of co-living. In light of the attendant irritation, however, the anticipation that algorithms will improve is shadowed by the possibility that perhaps they will never become sophisticated enough to offer the promised ease and convenience.

John Cheney-Lippold (2011) argues that algorithmic processes uphold "a new algorithmic identity" that promotes a shift from essentialist gender identity formation to a more flexible definition that detaches gender from its corporeal and societal arrangements. Yet Iida complains that she has been automatically approached with pregnancy testing promotions based on her gender and age, although she would probably rather see ads that follow an algorithmic logic that treats gender as "a vector, a completely digital and math-based association that defines the meaning of maleness, femaleness, or whatever other gender (or category) a marketer requires" (Cheney-Lippold, 2011, p. 170). If her online behavior were analyzed with more sophistication, Iida, not pleased with stereotypical baby reminders, could communicate this via her preferences for traveling and parties. Her irritation suggests that she expects the advertising machinery to recognize her better and generate an actual, personalized presence. Despite promises of a new algorithmic identity, the gender composition presented by advertising can still appear as a mere caricature, familiar from traditional advertising contexts.

There is nothing new in marketing and sales that promote normative gender differences, but targeted advertising makes stereotypes feel personally more insistent. Advertisements that replicate stereotypical gender categorizations point to the discrepancies between promises of fluid algorithmic identities and the rigid and crude ways in which automated advertising is actually employed. Ads that contain gender stereotypes appear backward in their prompting of fixed gender categories that many people would prefer to discard. The normative life course and heteronormative suggestions of procreation that ads strengthen are off-putting to young Finns who want to define their own aims, irrespective of customary expectations. In line with the notion of friction, however, their position in terms of automation is not one of resistance—they do not reject processes of datafication; rather, the irritation underlines that the machine is not reading them or their interpretation of what society should be in an accurate or sufficiently neutral manner. As Tsing (2005, p. 6) puts it, "Hegemony is made as well as unmade with friction."

The irritation with targeted advertising feeds into a wider argument about the mismatches between machinic categorizations and pursuits of selfhood and gender relations. The goal of activists aiming to influence current digital infrastructures is to resist unethical and discriminatory features of social sorting mechanisms and promote the deployment of subversive tactics to circumvent, manipulate, and disrupt algorithmic logics (Velkova & Kaun, 2021). Whereas the activist stance described by Julia Velkova and Anne Kaun (2021) promotes "algorithmic resistance," defined as a mode of political engagement arising "from alternative uses of platforms, in the aftermath of algorithmic logics," everyday discontent is less geared toward open resistance. Instead, personal experiences feature tensions that focus on how algorithmic operations, decisions, and choices fail to serve in the best possible manner. The daily discontents are felt in relation to the dominant structure of feeling and aspirations for smooth coevolution with algorithms.

By listening carefully to how irritation is articulated, we can begin to unpack the abstract claims made about algorithmic powers. Experiences of irritation offer concrete examples of how algorithms are seen to operate in the world, disclosing a love-hate relationship with digital technologies, a fitting description for interaction that is agonizing but impossible to relinquish. People use tracking devices to measure their stress and recovery, only to realize that the promise of well-being has turned into a nagging feeling of failure resulting from drinking an extra glass of wine when out with friends or not sleeping the recommended eight hours. Feelings of irritation triggered by mismatches with technologies can be emotionally more or less intense or turn into a more permanent frustration, depending on how people evaluate and judge their algorithmic companions.

Amplifying Stereotypes

Articulations of irritation repeatedly indicate discrepancies between how algorithms treat people and how people would like to be treated. Remember the frustration of the Danish man living in a same-sex marriage (discussed in chapter 1). He defined targeted advertising as "homo-spam" that treats sexuality as "a stand-alone classifier" for his identity. Rather than seeing the kind of human being he might be, the algorithm groups him stereotypically into a predefined category, homosexual, that becomes the overriding feature of his profile. By replicating stereotypes, the algorithmic logic operates as an amplifier of categorical differences, forcing people into groupings that are potentially discriminatory. Crudely targeted ads accelerate stereotypical classifications linked to gender and sexuality, but also race, ethnicity, and age. The irritation of our study's respondents signals that they feel that, societally, this a wrong move. As with heating up, which can be seen as an internalized zeitgeist of the critique of datafication, irritation with crude algorithmic mechanisms is an embodied response to the felt discrepancies between identity pursuits and algorithmic systems. Irritation is triggered by personal encounters with the algorithmic logic, yet it also points to sociocultural differences, as people in different contexts and places respond to and make sense of algorithm-related practices. Whether people are irritated with classifications that have to do with gender, race, sexuality, age, religion, wealth, or bodily appearance depends on the context, but the root cause of their emotional responses is shared.

Finland has an established history of gender equality politics (Holli, 2003), and both men and women are educated to spot biases in gender representations. It is thus no coincidence that gender is what people notice in their advertising encounters. The way Finns talk about how boring and infuriating it is that marketing replicates gender stereotypes highlights the friction between universally applied automation logic and the local characteristics of algorithmic encounters: women disapprove of pregnancy tests and wrinkle cream advertisements, men complain about being targeted by dating sites and claims of "hot singles near you." Iida, however, complains not merely about individual ads but also about the tendency to amplify and disseminate stereotypical gender separation with the aid of automated marketing. She is interested in technology, but the ads that she gets only feature "women's technology" (hair curlers and driers); she has also noticed that when her husband watches motorcycle videos, they come with odd pornographic anime ads. Following long-standing gender stereotypes, men and motorcycles are treated as a winning combination for selling sex, but digital marketing stereotypes are multiplying and appear to be everywhere. Here, algorithmic logic operates as a logic of repetition, ardently reviving and amplifying stereotypes that people have actively resisted and worked against for decades. The discontent and irritation reflects the feeling that, with algorithmic culture, societal developments are retrogressing, leading to more judgmental and rigid categories rather than moving forward into broadening tolerance and liberation. "Automation results in the abstraction of a task away from the motivations and intensions in which it is embedded," Mark Andrejevic (2019, p. 4) argues when describing how the road to automation paves the way to computational functions that can be used anywhere. Algorithmic logics are replicating stereotypes, but this in itself is not new. The newness that young women feel in their bodies and to which they respond with their irritation is related to the acceleration and amplification of stereotypes. Algorithmic techniques become a mechanism for scaling up age/gender categorizations. Baby-related advertising is everywhere.

Harshly Calibrated

Pregnancy test promotions remind young women that they are of childbearing age; wrinkle cream ads communicate that no woman over fifty should age without trying to act on it. The intimate nature of advertisements turns them into personalized reminders: you are aging, and you should do something about it. Whereas a successful encounter with an ad generates an enjoyable feeling of recognition—this product is exactly what I need—articulations of irritation tell the opposite story. They accentuate failures and vulnerabilities. The inability of algorithms to acknowledge the person in front of the screen is a reminder of the limitations of machines when it comes to truly recognizing who they are reaching. A vegetarian might be offered meat dishes; a meat lover is targeted with a vegan cookbook campaign.

The irritation with the machine's not seeing *me* is further connected to the slow and delayed manner in which algorithmic operations adapt to the flow of daily life. Personal experiences delineate the weaknesses of algorithms in the face of everyday routines and actions, which are never entirely foreseeable or complete, but always on the move. Here, irritation activates a broader conversation about how algorithmic systems overestimate predictability and underestimate changing circumstances. Kim presents a typical example. After an online search for Brixton hats, famous for their craftsmanship, every single advertisement that he notices a week later features them. There is no way to stop the stream of headwear. Algorithmic decisions are made based on data that is already old when represented to those whose actions it reflects; Kim's Brixton hat moment is long gone, but hats keep appearing in different colors and sizes. The stories told about haunting by advertising reflect a temporal lag that further connects to the felt inability of machines to distinguish between regular and rare occurrences. Teresa uses an anecdote about her boss to exemplify the problem. After buying a gift for his wife from an online store—a women's product that he did not care to specify—advertisements for similar women's products followed him like a shadow, although the recommendations did not correspond with his own profile in any way. Teresa points out how poorly developed digital advertising is if it cannot separate between typical and exceptional user behavior. The formal rules that the algorithmic operations follow lack contextual awareness, sometimes making automated suggestions appear harsh and discordant.

Ultimately, then, another major source of irritation is that although the aim of algorithms is to mirror probabilities of where people might like to go, they end up replicating where people have already been. Algorithmic systems fail to adapt to changes unless given a clear signal: if a middle-aged man changes his relationship status to single on Facebook, he starts to receive ads for dating sites. With the status change, algorithms are alerted that he might be in urgent need of female company. "And those ads do not remain in the realm of the neutral or societally acceptable," Pauli comments about the dating site marketing to which he has been exposed. Without clear indications, however, the ability of algorithmic systems to adapt to changing aims and desires remains arbitrary.

The irritation that stems from the ways that algorithms treat lived lives replicates the discomfort with commercial and administrative systems that try to get hold of people's lives. All classificatory schemes used for administrative and commercial purposes irritate, because they simplify and make legible much more varied and complex aspects of the everyday (Scott, 1998). The failures of algorithms to recognize people correctly frustrate in similar ways as the seeing functions of the state's administrative systems that endeavor to make its citizens decipherable. In terms of everyday lives, the seeing work of any top-down system is rigid and monolithic, rather than local, flexible, and divergent. What is new in the context of the emerging structure of irritation, however, is the immediacy of algorithmic relations. People feel algorithms as if they had the capacity to scan and search their lives in real time, and in the process, they learn about the partial and distorted ways that algorithms treat them. Algorithmic techniques that are used for sorting and scoring digital traces, then slotting and matching them for the purposes of advertising and sales, do not possess an enhanced capacity to peer comprehensively into lived lives, but the "seeing capabilities" are felt and observed by those whose data is being fed back to them as recommendations.

The more intrusive algorithmic systems become, the more closely they are involved in making deductions and decisions about people's lives. The discrepancies between machinic classifications and everyday aims are often suffered in silence; the algorithm is merely seen to be out of sync. Yet no matter where algorithmic processes take place, they share the quality of making judgments, even if machines do not truly have the capacity to judge. They produce algorithmic truths which, according to Louise Amoore (2020, p. 6), establish "patterns of good and bad, new thresholds of normality and abnormality, against which actions are calibrated." These calibrations are what cause people to react with irritation and frustration. The patterns and thresholds condense lives in ways that feel harsh and inadequate and call for more humane and varied algorithmic arrangements. The ultimate ideal and desire is for algorithmic operations to cause no tensions and conflict, but rather to align with how people see and feel about themselves.

Articulations of irritation resonate with the argument made in chapter 2: the more intimate the relations with digital technologies become, the more the machine is expected to behave like a friend or a buddy, sensitive enough to know whether suggesting the purchase of a pregnancy test or wrinkle cream is a good idea. Yet the machine frequently fails in this task because it is a machine and does not see like a human being. Algorithmic systems are unsuccessful at genuine recognition and personal treatment because they follow rules and procedures that do not care who the individual person is. They operate with the dehumanized dividual, computational traces assembled for the purposes of the algorithmic system. The system does what it has been programmed to do, deducing probabilities of features and actions, but it has no deeper interest in how the everyday develops or what is at stake. The irritation and frustration is caused by automation processes that operate with already known, normative, and judgmental categories to define lived lives, rather than with the felt and temporally changing contexts of daily events.

Shotgun Advertising

Discussing the advertising mismatches reported by other interviewees, Frank, the growth hacker—whose goal at work is to make digital marketing more effective—comments, "Algorithms are not developed enough." He readily admits that digital advertising lacks sophistication and works with oversimplified age-gender-location-based categorizations. Despite anticipation that algorithmic systems will work wonders, the implementations of such systems can leave a lot to be desired. Frank adds that advertising is hastily put together and with too little professional experience. Segmentation is easy to do on Facebook, but the precompiled segments might not work well in a small country like Finland. Frank describes how you can choose to target your advertisement to "parents of a newborn" or based on skin color, for instance, and, while he doubts the accuracy of this kind targeting, he knows that companies with insufficient marketing resources use such easily accessible demographic segments. Targeting based on skin color is illegal in Finland as it would count as ethnic profiling, but Frank claims that Facebook presents the option anyway, although he does not know whether anybody actually uses it. Nonetheless, segmentation tools encourage the setting apart of certain identifiable groups of people based on bodily characteristics, for instance, or significant life events, including marriage, birth of a child, or divorce.

Klaus, an engineer who works as a content strategist, adds that we often forget that automated marketing, and the digital marketing platforms that enable it, have mainly been built for the market in the United States with its very different volumes of data and users. He explains that when campaigns are carried out, it is vital to include data with "a wide enough net," meaning that the data set will inevitably contain data traces extracted from people who do not belong to the target group. The data sets covering Finns might be enriched with data traces of other Nordics—Swedes perhaps—or Estonians, producing situations in which the received recommendations can include suggestions that feel completely off target. Based on Facebook and other publicly available information, it is, for instance, difficult, if not impossible, to find precise information about Finns who might want to have children. The frequent irritation of young women with baby-related ads is a consequence of marketing that operates with as large a data set as possible, including the data traces of those who have chosen not to procreate. What is supposed to be targeted advertising ends up being random marketing, explaining the frustration of those who see the ads on their screens.

Here we are faced with a curious inequality in the global data world. Fewer than three million Finns regularly use Facebook, and Instagram has around two million users; currently, the fastest growing social medium is TikTok. What Klaus highlights is that the digital advertising world has its own centers and peripheries, and Finland is firmly on the periphery. As Sofia, a sales and marketing coordinator, puts it, "In a small population, advertising is always connected to the wrong people to some degree, because the reference groups cannot be narrowed down as precisely as they should."

Thus, the emerging structure of feeling, comprising frustration and irritation with algorithmic systems, is reinforced when digital services combine data traces in haphazard ways. Jimi, a student of computer science, compares the work of algorithms in advertising and social media to using a shotgun: precision is not a prominent quality. No wonder, then, that algorithmic recommendations feel out of sync, distorting how people want to interact and become enmeshed with algorithmic processes. Jenni moves the discussion to social media, directly associating the irritation that she repeatedly feels when exposed to "some useless person's stuff that I do not even want to follow" with the scatter of algorithmic targeting. She further stresses the point already made: algorithms react to life in ways that feel inadequate, unproductive, and call for improvements in their arrangements.

Algorithm Fatigue

Saara, a technology consultant, tells us how, sick with pneumonia, she was bedridden for three weeks. As she had no energy to watch anything that needed her full attention, she let "silly-silly" Netflix series run in the background as she slept and woke and slept again, finding the background noise of people's voices comforting. When she recovered, she wanted to get back to her normal viewing habits, but the "algorithmic bubble" had become so durable that she could no longer find anything meaningful to watch among the silly series she was offered. This inability of the recommender system to react to shifting conditions in daily lives concretizes their lack of contextual capabilities and responsiveness. The careful curation that is done to train and improve the recommender system can be ruined with an exceptional state of affairs, in this case illness, and there appears to be little one can do to avoid it. Despite predictive systems being described as proactive and preemptive, in the context of daily lives they can quickly lose their prophetic powers.

Saara knows enough about algorithms to understand that "personalization" is not actual personalization. Algorithmic classifications are not dealing with "Saara" but with features that define some aspects of people behaving like Saara. With three weeks of exclusively watching silly series, her reference group is "those who watch silly series." The ostensibly personalized dividual standing for Saara is constituted by combining computational features, including and excluding a variety of contexts. "Those who watch silly series" are not interested in art house movies or political satire.

Frustrations with recommender systems point to the impossibility of fixing a system that has been deranged by aberrant circumstances like Saara's illness—or Kari's road trip to a funeral with his grandmother, who wanted to listen to hymns. After that car ride, Kari kept getting hymn suggestions on Spotify that, he felt, had nothing to do with him. It was as if his grandmother had been added to his account as a user and he could not erase her. Veera, who works as the head of user experience in a technology company, connects her modified range of music recommendations to an incident some years earlier when her Spotify account was hacked, and the hacker's taste in Italian rap music ruined the system for her. Somehow, she never got back on track with her listening after that.

Professionals who feel let down by recommender systems due to "broken recommendations" or recurring algorithmic mishaps show symptoms of "algorithm fatigue." Ursula, who works as a project manager in an advertising agency, connects weariness with algorithms to time spent with inapt recommendations; she feels that the recommendations she gets are nothing but random combinations of "her own songs." To avoid the irritation, she would prefer more active engagement in filtering the content and choosing from given options. Yet she is given no such options, and the machine chooses for her, poorly. Kaius, a specialist in online communications, adds that developers of recommender systems appear intoxicated with the notion that they have the most complex and magically performing algorithms in the world. Yet he still gets by far the most valuable recommendations from his well-informed friends and acquaintances. The accumulated expertise that people have is not machine-readable information but might be based on a "gut feeling" that something is valuable or worth paying attention to.

Like many others, Veera, the head of user experience, misses recommendations compiled by humans, which feel novel and surprising, unlike algorithmic recommendations that reveal their genre too quickly. If one listens to "sighing female artists," that is all that one gets. Algorithm fatigue raises the question of whether personalization, as one of the main driving forces of the industry, might be an unproductive aspiration from the start. Here, the frustration with recommender systems resonates with the repeatedly posed question of whether technical expertise alone should drive the production of new technologies (Suchman, 2002). Despite companies investing heavily in recommender systems and adding new features to improve them, machinic references will never be as intuitive as people would like them to be, resulting in algorithm fatigue. When talking about their frustrations with recommender systems, people testify to the endless work that the machine requires. No matter how much they train the system, it appears to learn with great difficulty. As one of the interviewees puts it, "I can't be bothered to click no, no, no, hundreds of times."

Algorithm fatigue is a side effect of deepening algorithmic relations, suggesting that technological systems ignore the skills and know-how that people actually possess to intervene in and steer machinic processes. Recommender systems and digital assistants—because the same professionals also complain about assistants like Alexa—lack the ability to parse data in subtle and nuanced ways and fail to respond to ever-changing human experiences. Here the emerging structure of feeling, supporting irritation and frustration, underlines that algorithmic relations are first and foremost communicative relations. Algorithm fatigue is a consequence of the limited space for communication that current systems offer. Articulations of irritation describe how communication is blocked, expressing the difficulty of getting the message through to the recommender system. No matter how hard people try to maintain a two-way channel of communication—with Spotify, for example—the system fails to respond. While it is true that even if people were given the option to curate their responses to algorithmic recommendations they might not do that in the end, yet they would like to have the option. The felt discrepancies with recommender systems indicate an aspiration to communicate more swiftly with the machine.

What appears to be crucial is that automation should not erase possibilities of communication but rather strengthen them. Since algorithmic systems have limited capacity to address the contingencies of daily lives, people express a desire to communicate more freely with the system but also with other people, exercising their own judgment while doing so. The desire to communicate with algorithms reveals a longing for clever combinations of human and machine agencies, underlining the need to foster human qualities alongside algorithmic operations. In this view, the interviewees revisit the theme already discussed, suggesting that the collaboration of humans and machines should accentuate the strengths of both. Humans can only become more human with the aid of machines if they can express their aims and desires. Algorithm fatigue, however, highlights how machinic operations flatten human aims and devalue qualitative judgments, diminishing rather than augmenting processes of coevolving. Instead of reducing friction in human-machine relations, the lack of communication with machinic arrangements becomes an ongoing source of tensions. The unresponsive algorithm rubs us the wrong way.

In light of algorithm fatigue, a productive way forward would be to let go of the vision of the independent and all-knowing automated system and replace it with a more realistic notion that machinic agents have machinic qualities. In a study exploring the use of self-tracking devices to promote behavioral change in the context of life insurance (Tanninen et al., 2022b), current and potential life insurance policy holders expressed an interest in receiving real-life guidance from medical professionals, dieticians, and personal trainers, alongside the recommended use of the devices. For them, digital devices were merely a beginning in supporting relations, calling for human assistance. Evidence of frustrations with algorithmic relations supports the thesis that people do not want to be trapped in data loops and left alone with their devices; they also want expert advice and assistance. The aspirations to communicate with people alongside machines argue against the "digital solipsism" (Andrejevic 2019, 14) that moves the emphasis on sociality away from human communication. The more automation is used for erasing human resources, the more people can wish for open-ended communication, with devices and algorithms featuring as participants in interactions that include both humans and algorithmic agents.

Balancing Distorted Sociality

Articulations of frustration and irritation point toward forms of communication that are seen as lacking, biased, or misleading in light of everyday sociality. The distorting tendencies of digital services are a further source of frustration, as they favor commercially motivated ways of being in the world. When the frustration is located in the mismatches between social media and how they make people feel in their everyday lives, balancing acts are required, and the respondents in our study eagerly explain how they evaluate and deal with the distorting effects of social media. Max, an experienced publisher of vlogs and music, describes how his online profile is partly a marketing device, and he himself a brand. Curating his brand image requires ongoing attention; it is like trying to work on an appealing first impression, incessantly. The pleasures of digital engagements increase when Max feels in charge of how, and with whom, he communicates. Yet at times he thinks that it might be better simply to write down his ideas and keep them to himself, rather than sharing them with unknown strangers in digital space, where the distribution of messages and their interpretation cannot be fully monitored. On paper, he can hold onto his most intimate thoughts, without the fear that they will escape his control and become distorted by the reactions of unknown others who do not care about what he regards as intimate and worth protecting.

Anne, who works as a high school teacher, recognizes the self-branding that Max describes and cannot help questioning how distorting it is that digital spaces are more populated by self-promotion and commercially motivated messages than with actual "what I am thinking" posts. Facebook "has gone sour," she says, recalling that social media was a different place a decade ago, with people discussing their thoughts and whereabouts more openly. Aspects of lives were shared less consciously, while today young people practice similar curation in their posts as professional influencers. Facebook has changed the most, Anne thinks, but the drive to perform life and select only its top slices for online circulation has ruined the communal atmosphere of Instagram as well. For her, the balancing of interests has already failed in favor of commercial pursuits, and the algorithmic logic only strengthens this trend, as it favors the promotional.

Leo, who works at a cybersecurity company, feels no nostalgia for the early days of social media. Unlike Anne, he thinks that restricting information sharing to the most displayable aspects of daily lives is a healthy development, indicating that people are finally beginning to understand that "the first, last, and only control point" in terms of information sharing is whether to post or upload content at all. Consequently, it is a good thing that people carefully consider whether they should uncover details about their friendships, thoughts, feelings, beliefs, irrationalities, or aspirations. A lot of life is, and should be, left out of social media engagements. Leo adds that the narrowing of social media to marketing-friendly self-presentation is likely to make social media less appealing. He predicts that people will spend less time on social media, offering a welcome corrective to rampant data extractivism. For him, the most efficient way to combat the irritation and frustration caused by algorithmic systems is not to share.

Kira, in her mid-twenties, fails to see the development that Leo is describing. For her, social media not only distorts in terms of sociality but has an overall colonizing effect, as it is taken for granted that sociality is mediated by online exchanges. Kira refuses to download Snapchat and Instagram, adding, "It is perhaps a little rebellious, a small attempt to alert others that not everybody is there." She laments the assumption among her friends and acquaintances that social media updates are the sole source of information about each other's doings. Paradoxically, however, Kira keeps herself informed through her brother's and boyfriend's Instagram accounts; she cannot cut loose completely. It is of course somewhat ironic that she needs to use the accounts, but she fears that having her own would make her obsessed with the lives of others; she would lose focus and end up comparing her life to theirs. She feels that social media highlights only the shiny parts of life, distorting communication and ultimately misrepresenting life in general, but concedes, "When you look at the pictures, this is, however, difficult to keep in mind." As an antidote, she makes a practice of recalling an utterly boring party she once attended, which looked amazing in the pictures online. Kira has internalized a statement made by Annemarie Mol and John Law (2004, p. 57): "Keeping ourselves together is one of the tasks of life." To protect herself from representational distortions, she prefers face-to-face communication that offers a more rounded experience. She actively pushes back against the distorting tendencies of social media by meeting people in real-life contexts that are not defined by programmed sociality.

Sebastian adds that algorithmic logic teaches us to be "über-extroverted," always on display. As the über-extroverts flourish, the more silent or introverted can disappear altogether. Those who do not have the ability or courage to react quickly to others' posts with witty remarks are labeled stalkers or ghosts. The shy and still ways of being in the world are excluded, even ostracized. When Kim was serving in the military, obligatory for young men in Finland, he ended up turning into an online ghost. He had a lot of "empty time" that he filled with social media, browsing the posts of others and consuming their pleasurable experiences rather than partaking in his own. His mental health suffered, and he concluded that vicariously wasting time immersing himself in the good lives of others further weakened his emotional capacities. Eventually, he deleted his Facebook account.

My Data Benefits Others

"You think that you are using free services, but they are not free at all, even if it feels like it," Mia states. She is referring to one of the persistent annoyances with algorithmic systems, one that moves the conversation from the personal to the infrastructural level. The study of structures of feeling is able to address different registers, as the affective infrastructure comprises experiences that concern both the intimate and the public. Mia complains about the trade-off forced on her: she has to accept that data about her is gathered in exchange for the use of digital services. She finds it frustrating that the thousand or more data points on her characteristics, features, and actions are the currency with which to buy access to them. As currency, the data converts into a tradable object that can be "de- and re-aggregated, put up for auction, sold, remaindered, and reaggregated" (Skeggs, 2020, p. 733). As Beverley Skeggs reminds us, the automated trading of data traces happens incredibly fast; there may be tens of thousands of bids for Mia's data traces per second if they are seen as tradeable enough.

When thinking about the commercial webs and multiplying relationalities that are formed with the aid of her data, Mia is not sure if she wants to contribute to the current system with data traces at all. She has no desire to participate in mass consumption, prefers to buy her clothes used, and wears everything she buys for as long as possible. Yet while she tries to live in an anti-capitalist manner whenever possible in other spheres of her life, in the digital world that goal is impossible. Social media, integral to her way of being in the world, connects her inescapably to commercial networks. Just as personal experiences of fear indicate a collectively shared geography of fear, articulations of irritation push us to see informational asymmetries in a new light. Mia's frustration stems from the awareness that social media usage ties her to global power structures, sustaining capitalist values from which she is simultaneously trying to detach herself. She has no power over the commercial forces that make her a participant in value extraction, nor can she choose the kinds of companies who can benefit from her data.

Jimi is equally annoyed that the flows of data do not support his aims. Unlike Mia, however, he is not as bothered by the business models that rely on data gathering and auctioning as he is by whether he personally benefits from the data traces that he leaves behind. His thinking resonates with the Finnish-initiated MyData nonprofit, which is an advocate for a fairer data economy. Jimi would like to be able to control the flow of his data between different commercial agents, and he endorses the aim of MyData to promote the rights of individuals to become more active data citizens and data consumers by controlling the gathering, sharing, and analysis of personal data (Lehtiniemi & Ruckenstein, 2019). From Jimi's perspective, the goal of MyData is not to hinder the flow of data but to ensure the parallel advancement of processes and policies for protecting individuals' rights, while accommodating the industry's demands to process data in the development of innovative services. Jimi would like commercial agents of his choosing—Alexa or Siri, for example—to communicate actively with each other. He envisions how pleasant it would be if he merely had to tell the digital agent that he is bored for it to reserve him a ticket to a movie which, based on algorithmic profiling, would suit him perfectly.

In an ideal MyData world, the originators of data—those whose practices and behaviors contribute to data extraction—would be able to influence who gets value out of data and what kinds of algorithmic relations are supported. When people regard their data as a form of currency, their ideas echo the demands made by technology developers to combat data divides. Jaron Lanier's *Who Owns the Future?* (2013) argues that as commercial agents profit from digital traces, a portion of their gains should be distributed to the data subjects as remuneration for providing their data. In practice, this could be a form of data tax. Yet the gains that Finns expect in return for data extraction are typically not monetary. Jimi, for instance, would like algorithmic relationalities and resulting informational webs both to expand and to tighten so that the digital infrastructure supported him better, creating pampering bubbles. Rather than negotiating with his friends about what movie to see, he would be happy with the advice offered by an algorithmic buddy. The desire here is that algorithms maintain a silent infrastructure of communicative exchanges that would benefit the individual in whatever way fits the everyday.

The idea that personal data should benefit those who are its original contributors is reiterated in discussions among Finnish professionals, underlining their frustration with data power and the current informational asymmetry; this suggests that as data is used by various kinds of commercial agents, it matters where it goes and what is done with it. Yet Finns approach the question of data use in remarkably different ways, either criticizing commercial webs woven with their data traces or lamenting the fact that they are not sufficiently part of them. Here the concept of friction is useful for highlighting how processes of datafication are spreading with the aid of professionals convinced of their benefits, although their success remains partial and incomplete (Tsing, 2005). In Finland, the co-living with digital assistants that Jimi is after mainly appeals to younger, male technology professionals. The affective infrastructure reveals its gendered qualities, with the comfortability of young men to distribute their agency to data-driven systems. Others, such as Mia, would like to see personal data move less in commercial contexts; instead of espousing the MyData movement, she might like to join a NoData movement. The promotion of algorithmic systems, with their nudges and feedback loops, begs the question of whether such systems are increasingly subjecting individuals and societies to external powers, distorting lives and societal aims rather than enabling people to exercise their own judgment and interventional agencies. For Mia, the tightening digital infrastructure that the global business elite promotes feels suffocating rather than pampering.

Colonizing Forces in Context

The metaphor of data colonialism frames the frustrations of the algorithmic age. It resonates in everyday experience when people describe how data companies have the means to capture routine social acts and translate them into quantifiable data, to be analyzed and used for the generation of profit (Couldry & Mejias, 2019). The shared frustration here is that professionals in the fields of strategy, technology, and marketing are converting lives into digital data in order to colonize them. Once pronounced fit for companies, the data relations can be monitored, turning individuals, or rather dividuals, into salable and transferable data, which in turn can be used as material for targeted advertising, personalizing practices, and predictive analytics. Recommender systems and self-tracking devices quantify the self and others as they translate behavioral clues into data loops, while in exchange for the quantification of the everyday, people are consuming products and services based on data traces in new ways.

At the outset, data colonialism appears to be an overall global tendency. Yet locally its effects and implications remain manifold and ambivalent, underlining that despite the promotion of a global automation logic, algorithmic developments are unevenly distributed phenomena. Processes of datafication are steeped in history and depend on existing social stratification and oppression. Moreover, it is not clear what kind of colonial mechanisms are offered by data selection and data classification. Mikael, who approaches data colonialism through his studies in social anthropology, highlights the unknowns and uncertainties connected to the aims of the large data companies. While the notion of data colonialism suggests that commercial enterprises are on top of current developments, he suggests that this might not be the case, and that data companies might not be fully aware of how their actions affect people and societies. They pursue economic profit with the information that they have, but their doings have all sorts of intended and unintended effects, spilling over to trigger changes in everyday routines and practices. His suggestions raise the question of whether the data companies would operate as they do if they could fully grasp the scale of changes they are promoting. From the Finnish perspective, it might be difficult to fathom data companies' lack of societal consciousness, although it is possible to imagine them as the tobacco companies of our times: fully aware of the damage they are doing and hiding the consequences of their actions as tenaciously as possible, with the help of an army of lawyers. The uncertainty that Mikael introduces, however, suggests that algorithmic culture and processes of datafication are produced in interactions, in the friction that sustains unstable and unequal connections of power.

Whereas the notion of data colonialism argues that sensors and communication technologies colonize everyday life, extracting value by dispossessing individuals of their data, the Finnish perspective reminds us that digital colonization does not proceed without friction. People actively promote and work with the colonizing tendencies. As far as some algorithmic relations are concerned, people are very ready to become colonized: they build communicative relations with self-tracking devices and recommender systems and deepen algorithmic engagements by allocating decision making to automated processes. On the other hand, however, we see considerable frustration and pushback when algorithmic systems appear out of sync. Self-tracking devices, for example, are quickly abandoned if they start to distort everyday aims. The notion of friction aids in appreciating the processual nature of algorithmic relations and the many ways in which people coevolve in the course of taking part in them, situating our thinking about data and algorithms in the middle ground where technological and human agency condition one another; this helps to disengage us from linear projections of technologized futures. We can begin to see different versions of futures, including those that do not involve algorithms, if we learn to pose questions about the sensibilities and habits that come with algorithmic relations.

Welcome to the Global Living Lab

Klaus, the engineer who works as a content strategist, calls for greater societal regulation to govern companies that benefit from data extraction, claiming that all the wonderful products and services that we could build with algorithmic technologies will not materialize if we do not challenge informational asymmetries. The globally effective automation logic can be envisaged as a road forward, and the appeals for regulation reflect a desire for that road to be properly managed; speed limits and signs warning of bumps and curves along the way are required. Klaus says that the GDPR is an excellent start, but we need additional initiatives to promote responsible ways of using and distributing data; he mentions aviation as a historical precedent, with all its standards and implementation practices. The air industry operates within a global regulatory system wherein everyone follows the same rules, signs, messages, and techniques. Similar standardization is needed for the data-extracting economy to protect the interests of the society and consumers.

Klaus shares the opinion of many that algorithmic futures should not be steered only by market forces. In Europe, legal, regulatory, and ethical frameworks and new governance initiatives are promoted as a response to the proliferation of digital infrastructures and data-intensive automated services. Debates over privacy, fairness, transparency, and accountability are multiplying. The proposal for harmonized rules on AI in the European Union, for instance, attempts to boost the socioeconomic benefits and mitigate the harms related to algorithmic systems (European Commission, 2021). Indeed, ethical formulations depart from the notion that commonly shared values, ranging from solidarity and autonomy to trust and equality, are currently under threat (Prainsack & Van Hoyweghen, 2020; Sharon, 2018).

Regulation and ethical guidelines have traditionally been a way to strengthen the collective bases of society, and this is what current initiatives aim to do. Kennedy (2018, p. 24) points out the importance of seeing the ethical value in prospering in times of datafication *together*, especially "in light of the growing value attached to competitive individualism and neo-liberalism." Yet although regulation and ethical proposals are appreciated and much needed, they somehow talk past the concerns that people have, which is where they start to express doubts. Remember Henrik, who did not think that things would improve with informed consent. The geography of digital fear is a living testimony to failed protective measures. Even though policyholders of behavioral-based insurance, for instance, have signed informed consent forms, they are not sure what it means (Tanninen et al., 2022b). The multifaceted data relations constituted by behavioral policies remain a source of distrust and ambivalence. This finding stresses how regulatory approaches and supporting ethical initiatives might not respond to the unease with algorithmic technologies. Supporting this position, Sarah Pink and her colleagues argue (2022) that questions related to ethics are typically treated narrowly or at an abstract level, without paying enough attention to what people actually do and think. Ethical guidelines "take ethics out of the everyday."

The affective infrastructure of algorithmic culture calls for regulatory and ethical approaches that consider how people operate and feel in technologically mediated relations, but Leo, in his twenties, observes that experiential digital divides make this very difficult. He says that his parents' generation is seeking a moral high ground, yet with limited personal experience of the pervasive nature of digital technologies. Taking an all-knowing position in terms of the effects of digital technologies is not a good strategy, Leo thinks, because lives already lived are not the lives that people will be living. His thinking emphasizes that there can be no single version of what algorithmic culture does to us or our societies, highlighting the need to explore and assess, in careful detail, how algorithmic experiences are shaped within the larger context. Rather than regulation, Leo would like us to address the larger societal situation. Data relations are not like aviation; on the contrary, they are so complex that it is unrealistic to think about them as a stable object of regulation. What he is proposing is that regulating and managing such relations is a task that requires collaboration across differences. In order to create more profound knowledge about algorithmic futures, civil servants, company representatives, and ordinary people of different backgrounds and ages need to be part of the dialogue, as their various responses offer important clues to what needs improvement.

An Appeal for More Breathing Space

This chapter has engaged with the emerging structure of feeling that is sustained by articulations of frustration and irritation. Overall, irritation appears less consciously felt than fear, emphasizing that everyday actions are not a result of rational decision-making, but take place as bodies, devices, platforms, and infrastructures intersect with everyday practices, their histories, and future anticipations. What the irritation appears to be saying is that algorithmic systems cage people into unsatisfactory calibrations and user roles. People expect machinic classifications and algorithmic sorting to work silently in the background, sinking into the infrastructures of the everyday rather than offering a digital mirror that forces them to see themselves and their doings through computationally formed groups and identities that feel stereotypical, alienating, and erroneous. Personal stories address how algorithmic processes accelerate and amplify formulaic ways of grouping people: young women are seen as vessels of procreation, older women in need of fixing. Further frustrations concern distortions of sociality in social media, unequal data relations, and losses of digital sovereignty. Together, the experiences of irritation activate discontent on different scales and with different intensities. At the same time, however, they gesture toward a more overarching friction between globally promoted automation aims and local responses.

Exploration of irritation and frustration reframes the current debate by attending to the persistent tensions accompanying processes of datafication. As Ytre-Arne and Moe point out, perceiving algorithms as reductive does not merely refer to algorithmic biases; it also underlines that "human identity is too complex for the algorithm to understand" (Ytre-Arne & Moe, 2020, p. 14). The emerging structure of feeling, consisting of irritation and annoyance, points in two opposing directions: too much and too little algorithmic handling of life. Experientially, this means that successful living with algorithmic systems requires careful balancing acts and ongoing evaluation of whether such systems align with personal and societal aspirations. Thus, the focus on friction and associated discontents underlines the constant back and forth in algorithmic relations. The guarding of the limits of autonomy, for instance, begs the question of whether algorithmic systems are increasingly subjecting people to external powers, colonizing and distorting their lives rather than enabling them to exercise their own judgment and interventional agencies.

When personal experiences highlight how crude machinic logic fails to take into account the messiness of people's lives, their changing situations, circumstances, and aspirations, it signals an appeal for more breathing space in algorithmic relations. As algorithms are increasingly involved in forms of behavioral modification and people are pushed toward predefined actions and practices, the ensuing unease and irritation calls into question who defines the actions and practices. Here, the quest for breathing space addresses a desire to secure a space of reflection in the midst of the breathless chase after progress. Breathing space is needed because it allows us to reflect on and evaluate where we are going with algorithmic systems and how we are being transformed in the process. Ultimately, the wish to communicate with the machine reflects the desire to become involved in defining and steering algorithmic futures.

While algorithmic systems cannot live up to the expectations upheld by the optimistic anticipation of coevolving with machines, the need to include other perspectives than those of technology developers in algorithmic operations, in order to foster human agency and values, becomes an obvious step forward. Instead of maintaining the anticipation that technologies will become better, an alternative would be to let go of the anticipation and focus on the kinds of human-machine collaborations that are already out there. The emerging structure of feeling calls for taking irritation rather than anticipation as a starting point for thinking about future interactions between humans and machines, something that clashes with quasi-religious notions of algorithms as smart partners that will become even smarter. What the irritation calls for is the realization that humans are still smarter, and overriding their intelligence is not a very smart idea on any level.

5   *Care for Algorithmic Futures*

The preceding chapters have posed questions about who is guiding and controlling whom in algorithmic relations and what kinds of human-machine collaborations are emerging. Instead of focusing on how algorithms have powers over us—the question that occupies many researchers—I have introduced an alternative approach by illuminating emotional responses to what algorithms do or are thought to do. The algorithm has served as an entry point for exploring an affective infrastructure wherein algorithms are associated with actions and functions that are seen to impact us and our societies. Importantly, in daily experience the separation between the factual and the fabricated can melt away when algorithmic folklore becomes an integral part of the assessment of technical operations. The goal, then, has been to discuss how algorithmic agencies, powers, and collaborations are observed, felt, and lived with.

The aim of this concluding chapter is to outline what the discovered pleasures, fears, and irritations suggest in terms of coliving with algorithmic systems. Whether algorithms are seen as neutral, worth pursuing, scary, controlling, or rather avoided altogether depends on who is evaluating them, based on what kinds of criteria, and in which contexts. I revisit key findings concerning the three structures of feeling while taking advantage of a conceptual pair—the logic of choice and the logic of care—introduced by Annemarie Mol (2008) to discuss unifying themes that explain how individuals position themselves and others in relation to algorithmic systems. Mol's work took place in the health-care context, and applying her work to the feel of algorithms will give her concepts a new twist by redefining aspects of choice and care. In this context, care becomes evaluated favorably, as a positive social force, given that it is so obviously absent from the current arrangements around data and algorithms that sustain informational asymmetries. Data companies with profit-oriented motives might have no interest in the logic of care; they might even argue against it or try deliberatively breaking it. By demonstrating what it is, or could be, we can make the logic more visible and also argue against its absence.

As clarified later in the chapter, the logic of choice is supported by the dominant structure of feeling, which means that in order to take advantage of the logic of care and build on the oppositional and emerging structures of feeling when thinking about algorithmic relations, we need to first break out of the logic of choice framing. Since the logic of choice is currently the main mode for contemplating and practicing algorithmic relations, this is not an easy thing to do. When Finns imagine their lives with algorithmic systems they tend to reproduce the logic of choice. They treat algorithmic technologies as inevitable and suggest it is the task of individuals to learn new skills and not to fall behind.

The difficulty of imagining other possible future trajectories is precisely why engaging with the logic of care is so important. Examining the affective infrastructure in light of the two logics—choice and care—we begin to see what the different feelings suggest in terms of collectives and the political-economic conditions set by dataveillance and the intensifying logic of datafication. Importantly, the logic of care is not inherently superior to the logic of choice, and the two logics should not be seen as hierarchical or mutually exclusive. As I try to demonstrate, however, the logic of care has more potential when addressing collectively shared frustrations with algorithmic systems and the uncertainties and ambivalences involved. By attending to the frustrations, we can offer at least partial and tentative suggestions for how to promote communal and caring efforts to live well in a world shaped by both humans and algorithms. Approaching the future with an eye to both choice and care permits us to ask: What kinds of shared futures do we want? What forms of collective action do we need to instigate in order to get there? Instead of trying to predict or speculate on what will happen next, we can aim to uncover the potentials available to us in the here and now of the algorithmic age.

The Two Logics

By thinking in terms of logic, Mol references what is "appropriate or logical to do in some site or situation, and what is not" (Mol, 2008, p. 8). Logic refers to the coherence, not necessarily obvious to the people involved, implied in discourses, practices, and involvements with technologies. For instance, we tend to reproduce the logic of choice when making judgments regarding personal autonomy and empowerment. The logic comes into being, and can be identified, when individuals are seen as free to make choices when it comes to algorithmic systems. Technologies, in this case algorithmic technologies, are treated as neutral tools, as a means to an end (Mol, 2008, 57), becoming rooted and normalized through routine use. In light of the logic of choice, the adoption of technologies is a linear process. Once algorithmic systems are fully adopted, it becomes possible to evaluate their successes and make amendments accordingly (Mol, 2008, pp. 61–62).

The logic of care is a strikingly different mode of organizing practices, which becomes evident in how the two logics envision collectives forming. In terms of the logic of choice, a collective begins to form *after* people come together as capable individuals. This is how the technology and marketing professionals we interviewed tend to view current technological developments. Becoming literate and capable in algorithmic matters is an ethico-political ideal against which all people are measured. Liisa, who knows next to nothing about algorithmic systems, is expected to make similarly informed choices as Oskar, who is enthusiastic and skillful in his technology use. In light of the logic of care, however, the collective is not merely the product of aggregating capable individuals; rather, it starts from the notion that forming "a collective" requires that differences between groups be acknowledged and made visible (Mol, 2008, pp. 67–68). Therefore Liisa, with her limited technological capabilities, needs to be treated as a member of a different collective than Oskar. If Oskar wants to promote algorithmic literacy, for instance, he should not expect it to be of interest to Liisa, who might be committed to other concerns.

Whereas the logic of choice focuses on the choosing subject, the logic of care reaches beyond the individual and promotes open-ended processes that seek a desired result or outcome. Following the logic of care, the development work of an algorithmic system for supporting the well-being of young people, for instance, would not start from the predictive logic that separates out and targets youngsters at risk of being marginalized. The logic of care avoids promoting algorithmic truths that establish judgments about good and bad or normality and abnormality, instead aiming to create opportunities for all young people to improve their lives in association with others. Algorithmic relations, then, are not designed to reach certain predefined closure, such as determining a risk level by means of a scoring system; rather, the aim is to address young people, whether or not they have serious troubles or issues in their lives. When the algorithmic system reaches beyond merely targeting, scoring, and classifying, it opens communicative channels that support young people in their daily needs in an easily accessible manner, with the idea that this will ultimately lead to overall improvement in their well-being (Mol, 2008, p. 22). The goal is to figure out what is needed to care for youngsters in the best possible manner and, in the process, query what to do next (Mol, 2008, p. 93). Importantly, it might not even be of interest to decide beforehand whether a particular algorithmic setup is required to support the well-being of young people at all. The aim is not to promote technology but to consider whether technologies could be used to improve communication and collaboration in ways that support young people in a societally beneficial manner.

Equipped with the lens of care, we can approach algorithmic relations in a more open-ended manner when seeking to understand human-machine associations and related economic, political, regulatory, and cultural tensions. The logic of care can support movement in directions that are impossible to foresee in advance. Potentially, this can open up unexpected and perhaps unintended possibilities for technology development. Essentially, the logic of care does not start from what people know or want, but from what they need (Mol, 2008, p. 25). The logic of care sensitizes us to think about how technology developments could be made more responsive to the widely shared frustrations with them and aspirations to improve how they work (Ruckenstein & Turunen, 2020). When attention is directed toward unpacking people's needs and aspirations, we can begin to focus on those relations that should be formed or amended.

Crumbling Logic of Choice

The dominant structure of feeling, which sustains and is sustained by pleasurable algorithmic relations, aligns with the logic of choice by promoting the notion that people freely pick how they engage with technologies. They can select the best technology options available; if they rely on the latest applications and their security and privacy settings are up to date, they have no cause for distress. As the logic of choice positions individuals as masters and creators of their personal algorithmic journeys, it particularly appeals to those who feel, or would like to feel, that they have agency in algorithmic relations. The logic of choice becomes conspicuous when the professionals describe how algorithmic technologies create opportunities and open horizons of hope. Those most eager to endorse algorithmic futures are ready to learn new skills and promote practical and communicative engagements that generate conditions for improved practices. Importantly, they are also ready to work through the friction caused by the abrading qualities of automation logic on the contingencies of daily lives. Techno-optimists are confident that whatever bumps there might be ahead can be solved. "The problems with algorithmic systems need to be treated as pediatric diseases," as one of the interviewees put it. As the systems mature, they will be healed.

Yet thinking about the expansion and deepening of algorithmic relations, both outward to the globally wired data-extracting machinery and inward to the most intimate spheres of life, it is obvious that the digitalizing society treats people in an unequal manner. Even in a country like Finland, with its ideals of equality, people fall behind because they cannot access the digital society. If they do not have the personal ID code required for internet usage of a bank account, they cannot use many other services either. Yet this inequality consistently disappears from sight; the ethical guidelines that aim to steer algorithmic developments, for instance, do not typically address social stratification (Sloane, 2019). Here the focus on emotional responses is helpful, because the pleasures, fears, and frustrations that people share bring into view experiential digital divides that are neglected in the current debate. In our interviews, technology professionals recognize the problems with the logic of choice. They might think of their grandparents with a deep sense of empathy; they can see that grasping developments in technology does not come naturally to elderly people with slower cognitive skills, diminishing eyesight, or shaky hands. Others talk about children and young people who, despite their reputation as digital natives, are not as capable at managing technologies as we often imagine. They might be described as being at the mercy of the addictive and persuasive powers of social media, their sense of self distorted by the imperative to display only the shiny parts of their daily lives. Describing the self-promotion in which people engage, Anne, a high school teacher, referred to the current trend of self-branding, including performing life and selecting only the top slices of it for online circulation. She pointed out how hard young people have to work to keep themselves together in the midst of an online culture that distorts self-representations. To protect themselves, they might have to unlearn forms of programmed sociality that are characteristic of social media. What the interviewees appear to be saying is that even if the logic of choice remains an important ideal, holding onto genuine choices is becoming increasingly difficult, given the intimate nature of algorithmic relations and their tightening grasp on the self, sociality, and society.

Once we adopt algorithmic systems, there is no "outside" where choices about them can be made. As Langdon Winner (1980, p. 127) argues, "the greatest latitude of choice exists the very first time a particular instrument, system, or technique is introduced." People might, as individuals, refuse to adopt algorithmic systems, disconnect themselves from their technological companions, and ignore the hopeful horizons of automation; ultimately, however, it is increasingly difficult to operate in a digitalizing society without technologically mediated relations. The logic of choice crumbles when technical infrastructures offer little actual choice and sustain informational asymmetries that further weaken the potential to balance the advantages and disadvantages by way of individual choice. The more infrastructural digital technologies become, the less room there is to decide whether or not to engage in algorithmic relations. People are obliged to adopt technologies, as digital services mediate everyday communication and practices to such a degree that ignoring them is difficult, if not impossible. For instance, it might simply not be possible to refuse to participate in algorithmically mediated technology relations when making payments; getting insurance or financial aid; or becoming a student, a mother, or employee. Here, the logic of care works as an intervention, calling for a shift in perspective. If we want to maintain the option of informed choices, the logic of weakening choice, or no choice at all, should not be the starting point for algorithmic relations.

Aspirations That Matter

Feelings of inadequacy become observable in algorithm talk when people refer to their marginality and how they feel ostracized, lagging behind in digital developments. Algorithmic folklore informs us that common autodesignations when feeling ill-adjusted to digital society are "retiree" or "pensioner," somebody who has left active work in society. Young people might locate the nondigital or resistance to the digital in other times, the 1980s or the Stone Age. Mol points out (2008, p. 90) that the logic of choice is accompanied by guilt when people feel that they have not worked hard enough to keep up with developments in order to make informed choices; this manifests when people apologetically list features of technologies that they should have adopted. They also express the wish that they were more proactive in safety-related practices, readily acknowledging that they could do more to protect their privacy while realizing how easy it is to lose focus. At least occasionally, most people engage in practices that are not properly considered; they might download suspicious apps or links, use the same password for several services, share confidential information with inadequate protection, or send naked pictures of themselves to a stranger. There are countless opportunities to fail in privacy protection, and when people fail they know that they will be blamed for their mistaken actions. As internet users have the obligation to protect themselves, little empathy might be directed toward those whose information is hacked. As Teresa pointed out, in terms of digital harm, erroneously and ill-informed humans are the biggest problem.

Feelings of inadequacy are an outcome of the logic of choice, which treats individuals as responsible for installing and taking care of their own safety measures. Here the structure of feeling does not merely formulate an oppositional stance; it also offers a reformulation that gestures toward the logic of care. Since people are distressed by and uncomfortable with exactly the same aspects of routine digital technology use and remain suspicious of fast-paced technology development in strikingly similar ways, their emotional responses call for enhanced collective agency and protection in digital spaces. Thinking about the feel of algorithms through the care logic reaches beyond notions of individual agency when it stresses the necessity to provide mechanisms that are better tailored to people's own needs and aims. By following emotional responses, we begin to see how fears and frustrations are related to a shared difficulty with mastering algorithmic relations, suggesting that those relations would improve with the recognition of aspirations for different kinds of dealings with technologies, appropriately contextualized. As Annette Markham (2021, p. 400) puts it, "Aspiration functions as a navigational tool, through which people can chart their way out of a position of entrapment." Algorithmic systems promote feelings of being entrapped when they encroach on notions of privacy and autonomy and prevent people from deciding for themselves how they will participate and become involved.

Responding to discontents—whether connected to fears and frustrations or symptoms of algorithmic fatigue—requires technology-related practices that are designed to protect different kinds of human agency and their deficiencies. This is not, however, where work typically starts for profit-oriented technology companies, for whom "a one size fits all" approach is a more familiar and efficient way forward. The goal is to scale and not to diversify services to care for the needs and aspirations of those who are falling behind. The logic of care, then, argues against dismissing the sense of powerlessness and frustration that accompanies the expansion of the digital sphere. When the aim is to alleviate insecurities, one way forward lies in figuring out how professionals and nonprofessionals could collaborate in the longer term to improve the situation (Gangadharan & Niklas, 2019). Those who feel most exposed and insecure in digital spaces are rarely part of a conversation that would protect their rights and interests and aim to alleviate their insecurities. Taking the digital geography of fear as a starting point, for instance, would mean that all those who fear and distrust the digital world would have to be regarded as members of an important stakeholder group in service development. As a group, they could provide valuable information about the dimly lit parts of the digital and how to avoid their negative effects. As Mol (2008, p. 25) observes, "gathering knowledge is not a matter of providing better maps of reality, but of crafting more bearable ways for living with or in reality."

Participatory models for designing algorithmic systems aid in the effort of making different kinds of collectives visible by comparing insights, aspirations, and expertise. The interviewees in our study suggest some concrete steps in this direction when they express the desire to supervise algorithms alongside the designers to ensure that their voices are heard. They would like to evaluate the outcomes of the feedback loops containing the data about them to improve algorithmic operations. This is another kind of logic of choice, one that describes how algorithmic futures will depend on the choices and evaluations that we all make. In this format, the algorithmic system would not be taken for granted; instead, people could envisage how their feedback steers and updates decisions.

If needed, they could challenge the automated decisions made and call for revisions. Iida, for instance, pointed out how algorithmic logic revives stereotypical treatment that people have actively resisted. She is bothered by the dissemination of baby-related marketing, which appears to be everywhere. Despite assertions that digital technologies are forward looking in their aims, they become aligned with regressing societal developments, pulling us back to rigid heteronormativity. Iida's observations call for intervening in algorithmic accelerations that are replicating and scaling up stereotypes, as the machinic logic of repetition reproduces an outdated vision of gender.

Committing to the logic of care rather than the logic of choice is a commitment to think about the biases, insecurities, and inequalities of the digital environment. If indeed we are participants in an ongoing societal experiment, constituting a kind of global living lab, we need to take a proactive stance regarding the changes promoted. Acknowledging what bothers people aids in transforming individual frustration into a collective response that encourages institutional and systemic change. New kinds of sensors, devices, and services are constantly becoming available, and with each new algorithmic system, potential tensions and violations are introduced. Fears and insecurities cannot be amended with one-time technology fixes or regulatory measures; comprehensive societal efforts are required for things to improve. By reacting to the opaque nature of data companies, regulation can repair some of the damage that has been done, but in addition to that, we need to think carefully about the kind of society we are promoting. The ongoing testing and experimentation with algorithmic systems, in fields from health and education to advertising and security, means that the digital environment is very difficult to govern, and it is bound to generate new vulnerabilities. This suggests that we need to observe the changes in order not to miss intervention opportunities.

The testing and experimentation calls for changing the perspective on technologies. Steven Jackson (2014, pp. 221–222) advocates "broken world thinking" that allows us to see an "always-almost-falling-apart world," which is in a constant process of being fixed, reinvented, reconfigured, and reassembled into new combinations and possibilities. He argues that instead of exploring finished and newly emerging products of technology, we should highlight the unfinished and generative nature of sociotechnical systems and document processes of breakage, maintenance, repair, and renewal. The fragility of the always-almost-falling-apart world offers a more realistic way of approaching the algorithmic age in terms of the logic of care than the linear and product-oriented stories of technological progress.

Visibility of Feedback Loops

Collective participatory directions offer ways forward in the development work by acknowledging that in addition to experts who are developing and using algorithms professionally, those who are targets of algorithmic classifications have the knowledge to evaluate them. This kind of collective approach is, however, complicated by those algorithmic systems that make decisions in a way that is very hard to follow. In the face of such systems, both experts and nonexperts will have to think carefully about what it means to not know and whether "not knowing" is acceptable and legitimate in terms of future collectives. Without being asked about how current algorithmic relations could be improved, the people we spoke to repeatedly voiced calls for informational transparency with regard to algorithmic systems. Typically, they did this at the end of the interview, when identifying needed improvements. Appeals for transparency comprise a collective request to correct the dearth of information concerning algorithmic systems. People would like to know how data about them is gathered and profiles are compiled, whether their phones are listening to them or their data is being sold to third parties, and what happens to data traces in the process. In relation to informational actions on social media, they would like to understand why profiling measures are so clumsy, and why a particular like or other reaction, at a certain time, results in greater or lesser post visibility.

Transparency requests typically focus on the practices and organizing structures that define technology companies, rather than on the technical details of algorithms. The interviewees are well aware that their skills and competencies are not specialized enough to understand fully how algorithmic systems are built and operate. Since information on technical details could generate even less understanding and, consequently, heightened distrust in their capabilities to make informed choices, they are happy to leave the technical details to engineering teams. Markham (2021, p. 395) points out that "the disqualification of the general public" functions strategically to sustain the power of technology companies to shape our routines and social interactions. The tasks of creating the infrastructure are left to the experts, whereby the technologies themselves are treated as "unknowable."

Yet despite our interviewees' awareness that they lack expertise in matters of technology, they feel that increased transparency would provide them with greater knowledge and understanding of the modes of organization in algorithmic relations and the practices involved. Although we are connected to each other by way of data and algorithms, we are also alone with our devices, fed with information, terms, and conditions of use. Calls for transparency, then, pose questions about technically mediated relations that tend to disappear from sight when we become connected. Remember Mia's discomfort with the commercial networks that make her a participant in forms of capitalist value extraction that she would rather avoid. Her unease addresses the underlying economic arrangements that define platforms and infrastructures and force people to become data contributors in relations they have not chosen. The need to get a better grip on the algorithmic underlines the fact that the current nature of profit-driven data-related services is a pressing issue that calls for amendment.

Oskar, whose technology enthusiasm has become evident in previous chapters, wonders whether we should start talking about "an algorithmic milieu" that bundles together otherwise disconnected aspects of life. He envisions the algorithmic milieu with a poetic otherworldliness: "All of the invisible threads of code, threads of information. We can imagine how they are flying around us—a whole new level." This whole new level is revealed to us in partial and distorted ways. When people articulate responses to algorithms, they typically engage with processes of datafication one algorithmic relation at a time. What is harder to relate to is the systemic or infrastructural outlook that explains how algorithmic relations are linked to each other by way of processes of datafication. While the whole new level escapes scrutiny, we sense the invisible threads of code that are somehow connected to us, but we cannot really grasp what they do to us or society at large. In thinking about *my* algorithmic relations, the perspective remains haphazard and limited, while observations of algorithmic relations are further skewed because people tend to notice the exceptional rather than the routine. Emotional reactions triggered by algorithms can refer to failures and disconnects, moments when hidden infrastructures become newly visible. The failures might involve sloppy coding, poorly thought-out classifications, irrelevant recommendations, and key words that are completely off target. A focus on algorithmic relations adds experiential layers to technical and infrastructural failures, as they might be connected to identity pursuits and efforts to protect autonomy and desired forms of sociality. Indeed, emotionally charged personal reflections—feeling anxious, detached, or alone—that concern algorithms and digital services speak of the fragile nature of algorithmic relations.

Following the care logic, what people need is greater visibility of how they contribute to human-machine feedback loops with their own actions. They need a better infrastructural understanding of how they are knowingly and unwittingly involved in algorithmic relations. Thus, while appeals for transparency typically focus on how algorithmic systems work, they do not even begin to unravel the interconnections of such systems with everyday practices and aims. The notion of friction is useful here as it consistently draws attention to how people become complicit in shaping algorithmic systems when their practices strengthen the spread and profundity of the globally wired, data-extracting machinery. Algorithmic relations and related emotional responses materialize in the interaction and feedback loops of organizational practices and individual responses—continually forming, changing, and emerging as a result of human-machine interactions—and they cannot be programmed or fully mastered in advance. Both professionals and nonprofessionals familiarize themselves with processes of datafication in the course of their own experience, through short-term and habitual use but also as a result of improvisations, trials, failures, and misunderstandings.

Transparency requests continue to be voiced from the outsider position, one that offers firm ground from which to query political, economic, ethical, and regulatory developments, as well as claimed consequences for daily lives, although such requests rarely take into account the relentless feedback loops and how algorithms are bending us as we bend them. Paradoxically, the most vocal critics of the data machinery are often also avid users of social media services, indicating that critical decisions are not made as value judgments but result from practical activities, everyday doings that define the feedback loops of algorithmic culture.

Personal experiences reveal what different lives we can have with algorithmic systems, or even within the same systems, depending on what we do and our goals. Facebook, Instagram, Twitter, and Spotify serve people according to the kinds of algorithmic relations they promote with their own practices. We will not discover the ethical and political effects of algorithmic systems without paying careful attention to how they come into being when they participate in the daily flow of life. We need to understand, for instance, how the surveillant aspects of machines become justified when somebody else's Big Brother is treated as a caring "Dear Brother" (Siles et al., 2020, p. 6) who is observing you for a benign reason. Embracing the logic of care means that we need to familiarize ourselves with the human-machine combinations that are being promoted, while calls for transparency should embrace real-life examples, such as those discussed in previous chapters. These give clues to how algorithms work in different contexts, what they amplify in the process, and what kinds of consequences they have in terms of agentive abilities and societal developments.

Tensions with Autonomy

When people feel that they are no longer in charge of their technology relations, their sense of autonomy feels diminished. The powers of algorithms that push back against notions of autonomy become evident in talk about fears and frustrations that suggest that digital technologies encroach on the sense of self-determination. To fix the situation, users of digital services might try to reclaim their autonomy and detach themselves from unnecessary or harmful practices, temporarily if not in the long term, by deleting a social media account in an act of autonomy recovery. The ultimate indication of self-determination in relation to algorithmic services is the right to reject them; disconnecting has become a way to resolve tensions and diminish the frustration and annoyance felt about unsatisfying digital practices (Karppi, 2018).

The professional resources used for designing digital services, which take advantage of the tests and tools of behavioral psychology and behavioral economics, require a new kind of alertness from users of digital services, something that is observable in personal reflections on addiction and related loss of autonomy. When they talk about how digital technologies diminish their self-determination, the respondents of our study tend to reproduce the logic of choice; they rely in their accounts on the historically rooted notion of autonomy that has come to dominate the contemporary discourse on liberty, freedom of will, and self-determination (Taylor, 1992). In this perspective, autonomy is an entity that a person can "have" and technological systems can "control." Ella, for instance, talks about algorithms as hooking her; she can feel the power of technology companies guiding her actions. Yet while this clearly bounded notion of autonomy aids in critically positioning Ella in relation to algorithmic systems, it restricts the perspective to an either-or stance: one either thrives or fails in the autonomy game. When thriving, people might describe how they are on top of algorithmic systems and claim they are making their own judgments without algorithmic intervention. Mia, for instance, although frustrated with the way she promotes capitalist aims with her participation in algorithmic relations, still feels that she is above algorithms and can influence social media uses and content without being "determined" by them. The loss of autonomy, in contrast, is described as a surrender to the algorithmic logic, as becoming addicted.

Yet holding onto a clearly bounded notion of autonomy and protecting free will is increasingly difficult in the midst of algorithmic systems (Schüll, 2012; Sharon, 2017; Tanninen et al., 2022a). Rather than an entity that can be fully contained and protected, autonomy comes across as an everyday sensor that guides the evaluation of what is offered by algorithmic processes and whether they support personally and publicly valued goals. Here we need a relational understanding of autonomy that stems from situational understandings of values and feminist ethics (Mackenzie & Stoljar, 2000; Westlund, 2009) and treats autonomy as an active process, evaluated and weighed situationally in changing relations. In algorithmic relations, then, autonomy depends on the context; it is not a given reality but a relational quality that is actively both pushed against and worked on. For example, when autonomy has to do with pleasurable coevolving with technologies, close relations with machines inevitably lead to questions of self-surrender and addiction (Schüll, 2012). Schüll (2012) describes how machine gamblers in Las Vegas forget themselves while the experience of flow carries them forward; as they play, they are also "played by the machine." Similarly, personal experiences with recommender systems and self-tracking devices exemplify how one can overlook oneself when relying on the guidance of devices and services. People continue to use digital services, despite their addictive and surveilling tendencies and despite feeling ambivalent about the devilish ways in which technologies distort their selves and everyday lives, as long as they feel that they are getting something out of that relationship.

Distinguishing between beneficial and harmful aspects of algorithmic relations requires a perspective on autonomy that can deal with the current sociotechnical landscape. Articulations of emotions try to pin down how algorithmic relations enhance the sense of autonomy, while at other times the same relations feel distressing and even detrimental, as they amplify only certain aspects of the self and create technical lock-ins that feel reductive and limiting. Contextual aspects explain why a sleep-tracking application regarded by some as a fantastic booster for personal productivity goals seems repressive to others, as it turns sleep into a competence and a neoliberally driven activity. The differences between what is seen as convenient and worth pursuing in algorithmic relations and what turns out to be disturbing underline the importance of specific here-and-now contexts when assessing them, as a fine line can separate algorithmic relations that are self-enhancing and self-depreciating (Lomborg et al., 2018).

Thinking with the logic of care means that we need to grapple with this dual nature of autonomy. Acts of deliberately handing control over to an app or a service can be part of feeling self-directed; receiving demands and feedback from a technical system may feel motivating and uplifting if they confirm a sense of mastery and competence in that particular situation. In light of the logic of care, however, people also need support in detaching themselves from persuasive and stressful technology relations that appeal to addictive tendencies. Nobody can remain fully self-directed in a digital environment that is purposefully built to hook and trap users. Ideally, people need to feel that they can trust algorithmic relations to be on their side, even if they have distributed their decision-making power to an app or a service; in other words, they need to feel that algorithmic relations are not pushing them to make choices whose consequences they may later regret. Ultimately, emotional responses to algorithms suggest that pleasurable relations with technologies are maintained by respectful alignments with notions of autonomy. It is not always clear, however, what is respectful, suggesting that we need to keep assessing the intrusive nature of algorithmic services. If being connected means that one is alive, as one of the interviewees put it, the collective task is to ensure that we are alive in ways that make lives livable.

The push for self-chosen action mobilizes the quest for breathing space—a space wherein to foster goals, reasons, and self-definitions—an appeal inherent to emotional responses that are rooted in the tensions that define autonomy (Savolainen & Ruckenstein, 2022). The fear and frustration induced by algorithms has to do with the felt failure of algorithmic systems to respect us as self-authoring persons, while the notion of breathing space argues against behavioral modification efforts by making a distinction between calculable and traceable behavior and action that is endorsed by the self. Indeed, these reflexive qualities in particular, rather than manifest behaviors, are what define subjectivity and need to be protected. Rosa, for instance, finds the fact that people are unconsciously guided into certain modes of thinking an alarming aspect of digital services. The intrusive and addictive nature of digital services and the sense of threat to personal autonomy speak of the need to maintain a reflexive and autonomous self. Of course, commercial services have always tried to influence and persuade people, whether with advertising or service design, but algorithm-powered feedback loops further boost these aims. As with the repetitive logic of gender stereotypes, techniques of persuasion are extending their reach. Their traces are observable in tiny details like the notifications on the screens of our smartphones inviting us to return instantly to a service. The more people are pushed and jostled by algorithmic techniques, including behavioral modification tools, the more they need to think about questions of autonomy (Tanninen et al., 2022a). Since the logic of care does not start from what people want but what they need, one of the tasks of our times is to protect the breathing space that allows people to find their own ways and develop as reflective and autonomous selves.

Strengthening the Strengths of Humans and Machines

The quest for autonomy is connected to aspirations for improved flows of communication in algorithmic relations. Kasper, for instance, clicks on certain posts or news items to teach the algorithm how valuable he thinks they are, aiming to transfer his preferences to the machine. The active role of humans in algorithmic relations becomes visible when they participate in the work of training, curating, and fine-tuning algorithmic systems. When algorithmic relations are thought of as communicative relations, an unbroken flow of information is anticipated as people connect by way of their activities, and the algorithmic systems are expected to "listen" carefully. Here, the listening is not intrusive eavesdropping but part of the anticipation of frictionless co-living with algorithmic systems. In our material, the expectation of being heard is most comprehensive when people describe their coevolving with digital assistants and recommender systems. These systems should not merely pick up clues from earlier behavior; hopes of machinic companionship involve the desire for algorithms to adjust flawlessly to the changing flow of daily life. The machine's listening should translate into responsive and clever suggestions.

Machines are, however, not as responsive as people would like them to be, and in order to improve communication, the respondents of our study would like to step in to aid the recommender system. To avoid algorithm fatigue, they would like to see the "human touch" in algorithmic systems. The work of reviewing and modifying should be part of the interrelationality of humans and technologies, and market agents should take advantage of it in the design of their services. The fact that people would like to be able to signal changes in their circumstances, or to delete choices that they have made in atypical situations such as when sick or socializing, is an opportunity for service developers to learn more about their users. In seeking to develop perfect human-machine loops, they could allow people to do the fixing and adjusting to reduce friction and tie people more closely to machines. Yet failed communications speak of the tendency of algorithmic systems to ignore or reduce the participatory efforts of humans. People can of course participate, but in ways that align with the needs of technologies.

Otherwise the communication of needs and desires is typically not what drives technology development.

The providers of algorithmic services can downplay and disregard the human effort that goes into the design and implementation of such services, yet our interviewees are frustrated with the erasure of the human, arguing in different ways against the notion of machinic self-sufficiency and related tendencies to hide human efforts and excellence. Kaius, for instance, describes the recommendations of his friends and acquaintances as far superior to those of algorithms. Why not make human reviewers more visible in recommender systems? Others express a longing for algorithmic systems that would foster human agency alongside algorithmic operations; the collaboration of humans and machines should amplify the strengths of both. They would like to bring their own intelligence and judgment, currently not sufficiently recognized in the design of algorithmic systems, to the table, so that they could decide the degree to which algorithmic technologies become participants in shaping their lives and those of others.

If current algorithmic operations cage people into unsatisfactory user roles, thinking about better communication by cleverly combining human and machine strengths offers a possible way out. Here we are at the heart of the logic of care that promotes open-ended processes that seek a desired result or outcome. Uniting the best of machines and humans would require a careful fitting together of human aims and technical features; thus, the goal of automation should not be to erase human presence but to keep humans onboard, alongside the machines, if this promotes the preferred outcome. Human presence is of course not always needed, but when it is, it should be consistently developed and made visible. In terms of culturally rooted notions of the place of technology in history, involving and nurturing human strengths and efforts in technology development suggests a far-reaching change in perspective. We would have to think seriously about what we actually do for technologies and what we want them to do for us. Rather than algorithmic operations giving purpose and direction to human efforts, as Henrik, the life coach, suggests, we would have to develop our skills of anticipatory guidance in terms of the kinds of collective practices and visions we want to renew and promote. The magic of technologies would not be sustained by hiding human efforts; on the contrary, the efforts that go into the design, implementation, and use of algorithmic services would be celebrated as an integral part of the capacity of algorithms to open new paths and trails with their seeing and knowing capabilities. Ultimately, algorithms would see and know, not by themselves but because humans have equipped them with the skills to do so.

Seeds of New Paradigms

One of the unifying themes of the interviews has been the way the desire for better adjusted human-machine relations clashes with the crude ways that such systems currently deal with human aims and qualities. When algorithmic relations take hold of the everyday, aspects of life are modified by metric technologies. Despite promises of enhanced autonomy and improved skills, apps and devices can feel reductive and controlling, as they expect humans to operate in a predefined manner. This raises the broader question of whether preparations for the coming algorithmic age aim to create fitting digital environments for machines rather than humans. In discussing the experiences of new mothers, Helen Thornham (2019) describes how the apps they use create "data silences" when responding to breastfeeding practices; they care about the metrics—the frequency and length of feeds, for example—but not the cracking or bleeding of nipples or the pain involved. Despite promises that algorithmic technologies make us more human, we might end up feeling that our "experiences are devalued, unimportant and undermined" (Thornham 2019, p. 180) as qualities like empathy and context awareness are made redundant rather than enriched.

The power of anticipation flows through everyday discourse on technologies, building and reinforcing a hegemonic ideology of external power and control. When we criticize this ideology, we might end up strengthening it with an oppositional stance that creates "a discursive closure," feeding the feel of inevitability and powerlessness that shuts down options. "Alternatives are limited as we repeatedly tell ourselves and others that we have no control," as Markham (2021, p. 393) puts it. Responding to the current developments with the logic of care emphasizes, however, that we can find alternative ways forward. Stripping our lives of empathy and context and reducing us to human algorithms that mechanically assist machines, rather than being assisted by them, are not inevitable developments. We do not have to accept that we need to ready ourselves for machines, adjust our actions to technical processes, and become more machine-like in our behavior. The logic of choice is the dominant mode for practicing algorithmic relations, but it is not the only way to nurture and promote algorithmic relationalities.

Questioning the dominance of the logic of choice in algorithmic relations aids in breaking ground for thinking differently—and not only negatively—about the defining features of algorithmic systems. The dominant feel of algorithms is powerful, but it has gaps and breakages, and the emerging structure of feeling abrades the pleasures of algorithmic relations, reminding us of shortages and discontinuities in technical visions. Historically, hegemonic ways of seeing the world have been repeatedly disrupted and overturned; gradually, structures of feelings have changed over time. Since they contain both reproductive and resistant aims, feelings serve as potential points of defiance that make people rethink the taken-for-granted state of affairs. Technology professionals commit themselves to processes of datafication by imagining how things will improve with the aid of algorithmic operations. When frustrated, however, they might let go of the anticipation and focus on what is actually happening in the here and now. This shifts their public-facing enthusiasm into a more cynical stance, suggesting that their algorithmic worldview might be founded on empty promises; rather than opening new horizons, algorithmic systems may strengthen reductive tendencies, leaving little room for those less enthusiastic about the technology-driven perspective. The cynical remarks that these realizations trigger offer seeds for claiming back alternative futures that decenter the role of technologies as the focal point of future making.

Eeva, the crafts teacher, linked technological developments to ecological concerns, and if we had continued interviewing after 2020, we would most likely have heard ethico-political contemplation on the environmental consequences of datafication. Public discussion in Finland has started to connect questions of technological and economic growth to climate crises, suggesting that bringing environmental issues to the forefront of future visions creates a more trustworthy foundation for the connection between technology and future well-being. It is here that we might see a new paradigm forming. There is no doubt that algorithmic technologies can be used to solve environmental harms; at the same time, however, digital technologies accelerate problematic developments, as they have concrete environmental impacts in the form of the required raw materials, the electronic waste, and the energy required for computing capacity. Thus, in order to not remain trapped by the current feel of algorithms and the limits of current technology visions, locking us into a loop of existing narratives—"repeated every time we follow a news story on Twitter, stream a new series on Netflix, get caught up in the promise of the latest new tech development" (Markham 2021, p. 398)—we need to pay attention to what the oppositional and emerging structures of feeling are proposing. Thinking with the logic of care aids in this task by being responsive to aspirations to improve algorithmic systems, situate them in societally beneficial ways, and claim back alternative futures.

Ways Forward

I opened this book with a scene from a local community college in Helsinki after a talk I gave about algorithmic culture and related political economic processes. I had tried to convince the audience of the importance of studying everyday phenomena by connecting them to algorithmic technologies, but what one audience member at least was hoping for was help in navigating the digital environment. My first reaction, annoyance at her inability to see my point, situated us in different realms of algorithmic culture. I was the academic, charting the terrain of algorithmic relations, whereas she wanted to keep up with the relations that I was critically observing. After writing this book, I would no longer be irritated by her inability to relate to my goals. In fact, I would not even address the audience the same way I did, as it makes no sense to promote critical engagement with processes of datafication among people who are still feeling their way through the technology, without trying to tie current developments to their aims and concerns. If I were to redo my lecture, I would take advantage of personal stories about articulations of emotions to engage with political-economic processes in ways that are felt in the everyday. I would talk about the digital geography of fear and the pressures generated by the dominant feel of algorithms and describe the current lack of collective resources to build algorithmic services that align with the logic of care.

In pinning down personal and collectively shared feelings, algorithm talk seamlessly ties together the mundane and the infrastructural. The narrated emotional reactions are not simply individual responses; rather, they tell a more generalizable story of structures of feeling and related attempts to live well with algorithmic systems. The pleasures, fears, and frustrations come together to produce a blueprint of how such systems should be combined with human aims and efforts. Emotional reactions aid in recognizing troubling practices, but they also present alternatives that take advantage of the realm of possibilities. The current feel of algorithms suggests that human capacities—intuition, empathy, common sense, and contextual understanding—need fostering in the algorithmic era, and that emotional responses, irritation in particular, offer important signposting for developing improved algorithmic relations. The calls for human sociality, revealed by way of emotional responses, on the other hand, reveal individualizing and dehumanizing forces of algorithmic systems. The emotionally charged engagements with algorithms suggest that we need to actively protect the social fabric of the society.

Together, the chapters of this book argue that affectively charged technology relations signal a broader cultural shift that calls for our attention. We need more public discussion and research that starts with listening to and considering diverse voices and alternative perspectives on digital technologies. Experiences of fear and frustration expose variations in actual knowledge of algorithmic systems, but importantly, they also underline the fact that the sole authority to know algorithms should not be granted to experts. Attending to stories about algorithms and their effects offers the opportunity to include different kinds of collectives in the deliberation of algorithmic futures. Algorithmic folklore that is fed with everyday responses and misunderstandings is a means to address how top-down story lines of emerging technologies, but also their critiques, ignore and distort the way algorithms are experienced and lived with. Vernacular culture that develops in the absence of accurate information grows into an important resource for research as it sustains the feel of algorithms and influences future orientations toward what is at stake, worth sharing, or even believable.

Personal stories of algorithms describe in various ways how the current ways of handling algorithmic relations are not only societally insufficient but reproduce the individualistic tendencies of the logic of choice. The emphasis on privacy is particularly noteworthy in this regard. Despite repeated efforts to broaden notions of privacy—such as by introducing the concept of group privacy (Taylor et al., 2016)—it is still mainly treated as an individual pursuit. Individual acts of withholding or curating information to enhance privacy and evade precise profiling are a means to speak back to a system that feels invasive and untrustworthy. The collective discomfort will not, however, be solved by individual privacy protection and associated regulation of data movements. Living with algorithmic systems calls for more responsive and comprehensive ethical and policy approaches, as well as ongoing assessment of options allowing responsible and informed choices to be made collectively as a society. Data and algorithm literacy programs are important and well-intentioned, but their approach tends largely to focus on technical knowledge and skills, which is merely a starting point when posing questions related to livability in the midst of emerging technologies. By seeking a more thorough understanding of everyday practices and related emotional responses, we can ask specific questions about what is at stake in decisions affected by algorithms, given that both the logic of choice and the logic of care are present and at times coexist when people make sense of and pursue algorithmic relations.

Due to the complex nature of such relations, we will continue to hear affectively charged assessments of the impact of algorithmic systems on individuals, communities, and societies. With its ability to deal with contradictions and ambivalences, the focus on the affective infrastructure of algorithmic culture is helpful in this regard. The study of structures of feelings opens a novel perspective by addressing agencies and uncertainties, strengths and vulnerabilities in algorithmic relations. When dealing with everyday discomfort, for instance, the goal is not merely to analyze how people feel and respond to algorithms, but to use the discomfort to expand the existing domain of social critique. By following how people feel about algorithmic agencies and how they position themselves in relation to them, we learn about the persistent tensions raised by everyday technology uses. The desire for autonomy, for instance, suggests that a fine line can separate the acceptable and the unacceptable in algorithmic relations.

The metaphor of the breathing space, which speaks to the experience of being relentlessly chased by algorithms that are trying to steer one's actions in ways that might not be self-chosen, emerges as a powerful sign of our times. The persuasive nature of digital services and the sense of the threat they pose to autonomy strengthen the need to maintain a reflexive and autonomous self. Thus, the quest for breathing space is a logical reaction to intrusive algorithmic powers: technologies push us to think about questions of autonomy. And when we think, we also feel. The critique of current technologies might be somatic: the voice of discomfort rather than well-formed arguments and public statements. Yet it is critique all the same. People know, in their bodies, that the dataveillance and continuous tracking of their everyday actions is not right. The distress, described in Finland as heating up, indicates an embodied awareness of the negative effects of processes of datafication and communicates frustration that so little can be done to diminish them in order to improve the current situation.

In light of the structures of feeling discussed in this book, then, the answer to the question of how to live alongside algorithmic agencies is rooted in different algorithmic relations and changing situations and circumstances; there are many ways to live well with machines and also to suffer as a result of them. Since the empirical material used here has been gathered in Finland, the findings are biased toward the aims and concerns of residents in a Nordic welfare society. Finns enjoy a high level of public trust, and they are protected from the most detrimental forms of surveillance enabled by current sensor technologies. Their highly digitalized society actively mitigates the harms and risks connected to algorithmic systems and is therefore well positioned to pursue models that combine algorithmic technologies with human capabilities. Yet there is still a widely shared unease in Finland that processes of datafication threaten core values in society, including privacy, equality, openness, autonomy, and trust. The companies that offer algorithmically powered consumer services—thus defining the norms, rules, and values for how algorithmic procedures should play out in the everyday—promote suggestions for what the algorithmic era should look like that are in tension with Nordic welfare society ideals. The tendency of digital technologies to make some people flourish while others feel that their agency is reduced becomes visible in emotional responses to algorithms, with those who do well with technologies approaching them with enthusiasm while those feeling vulnerable in relation to current digital developments are more likely to voice concerns and fears.

I have argued that emotionally charged engagements with algorithms challenge us to think about the kind of society we want to live in and who we want to become in the process. In order to find productive ways forward, we need to reach beyond the anticipatory tendencies that fix our thinking about technologies in a loop of existing narratives. The common way of approaching technologies as either an opportunity or a threat stands in the way of finding alternatives. Feelings continue to be a valuable tool of navigation in this regard, because they are triggered by both pleasures and discontents, which can help to delineate problematic practices related to the algorithmic, meanwhile calling for alternatives that are currently ignored or suppressed. The irritation and frustration that constitute the emerging structure of feeling open up opportunities to see how the more persistent nature of technologies rubs us the wrong way. As a cultural pattern, irritation is somewhat difficult to unpack, but if we follow its resistant qualities carefully, we can see it as an agent of history that calls for our attention. What the irritation appears to be stressing is that there is an urgent need not only to voice concerns and emotional responses that are neglected in the current debate, but also to create mechanisms for bringing such voices together as a collective. In terms of protecting shared values, we need to demand more as consumers and citizens from digital services. Suggestively, irritation brings us closer to actual algorithmic developments, making us more attentive to the current feel of algorithms and the different feelings involved in creating and reproducing algorithmic relations. Yet as I have also highlighted, algorithmic relations, many of which are naturalized to a degree that makes them difficult to observe, do not necessarily trigger any particular emotions. The neutral feel of algorithms keeps current informational infrastructures and related power imbalances in place. Thus, while the focus on emotional responses might exaggerate or conceal some aspects of algorithmic culture, it also makes its contextual nature tangible. Pleasures and frustrations underline that technologies do not stand on their own but are deeply ingrained in everyday visions, practices, and ideologies; for example, it is no coincidence that questions related to autonomy and informational openness bother Finns, as these are publicly cherished values that they have learned to expect in life.

In thinking about future alternatives, questions related to shared goals continue to signpost ways of appreciating and organizing algorithmic systems. Finnish experiences tend to accentuate discomfort with commercial services and their algorithmic logic rather than public sector uses of automated services, raising questions about how generalizable this finding is. It is evident that the feel of algorithms is tied to the ways people perceive the relationship between the citizen and the state. Finns mainly think of the state as a benevolent actor in their lives, with caring surveilling tendencies, but this is of course far from universal. Comparing local responses to algorithmic systems—ranging from targeted advertising and credit scoring to border control and police risk assessment algorithms—would shed further light on the affective infrastructure of algorithmic culture. Whereas border control and predictive policing tend to target people within a clearly defined spatial and temporal realm, commercially applied algorithmic techniques aim to pamper consumers with personalized suggestions all the time and regardless of their location. While the first approach is exclusionary and risk based, the second is seductive and expansive, focusing on creating a tailored informational space—a comforting and caring bubble—for the consumer. Algorithmic technologies create new scales, spaces, and temporalities to be accounted for and ruled by means of automation (Masso & Kasapoglu, 2020), triggering emotional responses. Comparative approaches in different contexts and locations would deepen understandings of the alignments and mismatches between digital technologies and everyday lives, while also clarifying how emotional responses to algorithms mediate personal experiences of vulnerability and political-economic processes of datafication. The dehumanizing features of algorithmic systems, for instance, need to be understood in an empirically grounded manner so that we can better identify them and detach ourselves from their harmful elements.

The exploration of the affective infrastructure of algorithmic culture promotes a deeper appreciation of a rather obvious point: we are not, and will not become, machines. The emphasis on humanness in relation to machines is not new, but it becomes newly manifest with current algorithmic systems that press against what people imagine to be human. The fear and discontent that accompany current technology arrangements merge into a very simple suggestion: people need to be heard and their daily lives need to be taken into account when visions of the algorithmic age are presented. In order to live well, or better, with algorithmic systems, we need a realistic understanding of their capabilities, but we also need a more thorough understanding of human aims and agencies in the midst of current developments. Algorithmic futures do not merely happen but require constant effort to become what they will be. The most enthusiastic proponents of algorithmic futures tend to be young men, which suggests that, in addition to being socialized into the ideal that it is the potential of algorithmic operations that gives purpose and direction to future efforts, their positive algorithmic outlook maintains a gendered split in algorithmic culture. Even if we tend to associate technical efforts with mathematical calculation and rationality, devoid of feminine emotionality, we are confronted with an affective realm of masculinity that defines what is worth pursuing in being and becoming a human. This is connected to a comfortability with digital solipsism, promoting programmed interactions with digital devices and assistants and moving the emphasis away from human communication. Rather than communicating with other people, young techno-optimists are comfortable talking to digital assistants, expecting them to learn to read their minds with their computational capabilities.

When the machine turns into a mirror that teaches us what is human or not human, it gives us the incentive to protect human qualities and sociality that machines will never have. Whereas techno-optimists might live with the anticipation that machines will eventually become like fellow humans, the everyday feel of algorithms is characterized by a longing for actual human touch. The limited nature of communication with machinic agents activates hopes that human choices and expertise become acknowledged in algorithmic systems. Despite advances in computational technologies, machines will never be able to feel the enjoyment, frustrations, and irritations that have produced the material for this book. While the messiness of people's lives—the changing situations, circumstances, and aspirations—can appear to be a problem for the developers of algorithmic systems in that the human is the most likely cause of system error, these characteristics make sure that we will not become "uniform, averaged, smoothed out: persons without qualities" (Davis & Scherz, 2019, p. xxxvi). Irritation and frustration should be treated with respect, as they continue to push back on the application of standards and forms of behavioral modification. Future critical research into algorithms and datafication should not bypass the irritation that people feel, but use it to foster care in our collective efforts to live well in an algorithmic culture.

References

Abidin, C. (2016). Visibility labour: Engaging with influencers' fashion brands and #OOTD advertorial campaigns on Instagram. *Media International Australia, 161*(1), 86–100.

Adams, V., Murphy, M., & Clarke, A. (2009). Anticipation: Technoscience, life, affect, temporality. *Subjectivity, 28*(1), 246–265.

Ahmed, S. (2004). Collective feelings. *Theory, Culture & Society, 21*(2), 25–42.

Amoore, L. (2020). *Cloud ethics.* Durham, NC: Duke University Press.

Andrejevic, M. (2014). The big data divide. *International Journal of Communication, 8*(1), 1673–1689.

Andrejevic, M. (2019). *Automated media.* New York: Routledge.

Aoun, J. (2017). *Robot proof: Higher education in the age of artificial intelligence.* Cambridge, MA: MIT Press.

Arvidsson, A. (2005). Brands. *Journal of Consumer Culture, 5*(2), 235–258.

Bassett, C., Kember, S., & O'Riordan, K. (2020). *Furious: Technological feminism and digital futures.* London: Pluto Press.

Bateson, G. (1936). *Naven: A survey of the problems suggested by a composite picture of the culture of a New Guinea tribe from three points of view.* Cambridge: Cambridge University Press.

Baym, N. K. (2013). Data not seen: The uses and shortcomings of social media metrics. *First Monday, 18*(10). https://firstmonday.org/article/view/4873/3752

Baym, N. K. (2015). *Personal connections in the digital age.* Cambridge, UK: Polity.

Benedict, R. (2019). *Patterns of culture*. London: Routledge.

Berg, M. (2017). Making sense with sensors: Self-tracking and the temporalities of wellbeing. *Digital Health*, *3*, 205520761769976.

Bericat, E. (2015). The sociology of emotions: Four decades of progress. *Current Sociology*, *64*(3), 491–513.

Bishop, S. (2019). Managing visibility on YouTube through algorithmic gossip. *New Media & Society*, *21*(11–12), 2589–2606.

Brembeck, H. (2008). Inscribing Nordic childhoods at McDonalds. In M. Gutman & N. de Coninck-Smith (Eds.), *Designing modern childhoods: History, space, and the material culture of childhood* (pp. 269–281). New Brunswick, NJ: Rutgers University Press.

Bruns, A. (2008). *Blogs, Wikipedia, second life, and beyond: From production to produsage*. New York: Peter Lang.

Bruns, A. (2019). *Are filter bubbles real?* Cambridge, UK: Polity Press.

Bucher, T. (2013). The friendship assemblage: Investigating programmed sociality on Facebook. *Television & New Media*, *14*(6), 479–493.

Bucher, T. (2016). Neither black nor box: Ways of knowing algorithms. In S. Kubitschko & A. Kaun (Eds.), *Innovative methods in media and communication research* (pp. 81–98). Charm, Switzerland: Palgrave Macmillan.

Bucher, T. (2017). The algorithmic imaginary: Exploring the ordinary affects of Facebook algorithms. *Information, Communication & Society*, *20*(1), 30–44. https://www.tandfonline.com/doi/full/10.1080/1369118X.2016.1154086

Bucher, T. (2018). *If . . . then: Algorithmic power and politics*. New York: Oxford University Press.

Burrell, J. (2016). How the machine "thinks": Understanding opacity in machine learning algorithms. *Big Data & Society*, *3*(1), 205395171562251.

Cheney-Lippold, J. (2011). A new algorithmic identity: Soft biopolitics and the modulation of control. *Theory, Culture & Society*, *28*(6), 164–181.

Cheney-Lippold, J. (2017). *We are data*. New York: New York University Press.

Christin, A. (2020). *Metrics at work: Journalism and the contested meaning of algorithms*. Princeton, NJ: Princeton University Press.

Cook, D. (2004). *The commodification of childhood: The children's clothing industry and the rise of the child consumer*. Durham, NC: Duke University Press.

Cotter, K. (2019). Playing the visibility game: How digital influencers and algorithms negotiate influence on Instagram., *New Media & Society*, 21(4), 895–913.

Couldry, N., & Mejias, U. A. (2019). *The costs of connection: How data is colonizing human life and appropriating it for capital.* Stanford, CA: Stanford University Press.

Davis, J. E., & Scherz, P. (2019). Persons without qualities: Algorithms, AI, and the reshaping of ourselves. *Social Research: An International Quarterly*, 86(4), xxxiii–xxxix.

Deleuze, G. (1992). Postscript on the societies of control. *October*, 59, 3–7.

Dencik, L., & Kaun, A. (2020). Datafication and the welfare state. *Global Perspectives*, 1(1). https://online.ucpress.edu/gp/article-abstract/1/1/12912/110743/Datafication-and-the-Welfare-State?redirectedFrom=fulltext.

Draper, N. A. (2017). From privacy pragmatist to privacy resigned: Challenging narratives of rational choice in digital privacy debates. *Policy & Internet*, 9(2), 232–251.

Draper, N. A., & Turow, J. (2019). The corporate cultivation of digital resignation. *New Media & Society*, 21(8), 1824–1839.

DuFault, B. L., & Schouten, J. W. (2020). Self-quantification and the datapreneurial consumer identity. *Consumption Markets & Culture*, 23(3), 290–316.

Edwards, J., Harvey, P., & Wade, P. (2010). Technologized images, technologized bodies. In J. Edwards, P. Harvey, & P. Wade (Eds.), *Technologized images, technologized bodies* (pp.1–35). New York: Berghahn Books.

Eslami, M., Rickman, A., Vaccaro, K., et al. (2015). "I always assumed that I wasn't really that close to [her]": Reasoning about invisible algorithms in news feeds. In *CHI '15: Proceedings of the 33rd annual ACM conference on human factors in computing systems* (pp. 153–162). New York: ACM. https://dl.acm.org/doi/10.1145/2702123.2702556

Espeland, W. N., & Stevens, M. L. (2008). A sociology of quantification. *European Journal of Sociology*, 49(3), 401–436.

Eubanks, V. (2018). *Automating inequality: How high-tech tools profile, police and punish the poor.* New York: St. Martin's Press.

European Commission. (2021). Proposal for a regulation laying down harmonised rules on artificial intelligence (Artificial Intelligence Act). 2021/0106/COD.

Fourcade, M., & Healy, K. (2017). Seeing like a market. *Socio-Economic Review, 15*(1), 9–29.

Fors, V., Pink, S., Berg, M., & O'Dell, T. (2020). *Imagining personal data: Experiences of self-tracking.* London: Routledge.

Frau-Meigs, D. (2000). A cultural project based on multiple temporary consensus: Identity and community in Wired. *New Media & Society, 2*(2), 227–244.

Gangadharan, S. P., & Niklas, J. (2019). Decentering technology in discourse on discrimination. *Information, Communication & Society, 22*(7), 882–899.

Gerlitz, C., & Helmond, A. (2013). The like economy: Social buttons and the data-intensive web. *New Media & Society, 15*(8), 1348–1365.

Gillespie, T. (2013). The relevance of algorithm. In T. Gillespie, P. J. Boczkowski, & K. A. Foot (Eds.), *Media technologies: Essays on communication, materiality, and society* (pp. 167–194). Cambridge, MA: MIT Press.

Gillespie, T. (2016). Algorithm. In B. Peters (Ed.), *Digital keywords: A vocabulary of information society and culture* (pp. 18–30). Princeton, NJ: Princeton University Press.

Goldstein, D. (2015). Vernacular turns: Narrative, local knowledge, and the changed context of folklore. *The Journal of American Folklore, 128*(508), 125–145.

Haapoja, J., Laaksonen, S. M., & Lampinen, A. (2020). Gaming algorithmic hate-speech detection: Stakes, parties, and moves. *Social Media & Society, 6*(2), 205630512092477.

Haapoja, J., Lampinen, A., & Vesala, K. M. (2021). Personalised services in social situations. *Proceedings of the ACM on Human-Computer Interaction, 4*(CSCW3), 1–21.

Haraway, D. (1998). The biopolitics of postmodern bodies: Determinations of self in immune system discourse. In S. Lindenbaum & M. M. Lock (Eds.), *Knowledge, power, and practice: The anthropology of medicine and everyday life* (pp. 364–410). Berkeley: University of California Press.

Hargittai, E.. & Marwick, A. (2016). "What can I really do?" Explaining the privacy paradox with online apathy. *International Journal of Communication, 10,* 3737–3757.

Hayles, N. K. (2006). Unfinished work: From cyborg to cognisphere. *Theory, Culture & Society, 23*(7–8), 159–166.

Highmore, B. (2016). Formations of feelings, constellations of things. *Cultural Studies Review*, 22(1), 144–167.

Hochschild, A. R. (1983). *The managed heart: Commercialization of human feeling*. Berkeley: University of California Press.

Hochschild, A. R. (2002). The sociology of emotion as a way of seeing. In G. Bendelow & S. J. Williams (Eds.), *Emotions in social life* (pp. 31–44). London: Routledge.

Hogan, M. (2015). Data flows and water woes: The Utah Data Center. *Big Data & Society*, 2(2), 205395171559242.

Holli, A. M. (2003). *Discourse and politics for gender equality in late twentieth century Finland*. Helsinki: University of Helsinki.

Jackson, S. J. (2014). Rethinking repair. In T. Gillespie, P. J. Boczkowski, & K. A. Foot (Eds.), *Media technologies: Essays on communication, materiality, and society* (pp. 221–240). Cambridge, MA: MIT Press.

Karakayali, N., Kostem, B., & Galip, I. (2018). Recommendation systems as technologies of the self: Algorithmic control and the formation of music taste. *Theory, Culture & Society*, 35(2), 3–24.

Karppi, T. (2018). *Disconnect: Facebook's affective bonds*. Minneapolis: University of Minnesota Press.

Karppi, T., & Crawford, K. (2016). Social media, financial algorithms and the hack crash. *Theory, Culture & Society*, 33(1), 73–92.

Kennedy, H. (2018). Living with data: Aligning data studies and data activism through a focus on everyday experiences of datafication. *Krisis: Journal for Contemporary Philosophy*, 38(1), 18–30.

Kennedy, H., & Hill, R. L. (2017). The feeling of numbers: Emotions in everyday engagements with data and their visualisation. *Sociology*, 52(4), 830–848.

Koskela, H. (1997). "Bold walk and breakings": Women's spatial confidence versus fear of violence. *Gender, Place & Culture*, 4(3), 301–320.

Kristensen, D. B., & Ruckenstein, M. (2018). Co-evolving with self-tracking technologies. *New Media & Society*, 20(10), 3624–3640.

Lanier, J. (2013). *Who owns the future?* New York: Simon & Schuster.

Larkin, B. (2013). The politics and poetics of infrastructure. *Annual Review of Anthropology*, 42(1), 327–343.

Latour, B. (2005). *Reassembling the social: An introduction to actor-network-theory*. Oxford: Oxford University Press.

Lehtiniemi, T., & Ruckenstein, M. (2019). The social imaginaries of data activism. *Big Data & Society*, 6(1), 2053951718821146.

Lomborg, S., & Frandsen, K. (2016). Self-tracking as communication. *Information, Communication & Society*, 19(7), 1015–1027.

Lomborg, S., & Kapsch, P. H. (2020). Decoding algorithms. *Media, Culture & Society*, 42(5), 745–761.

Lomborg, S., Thylstrup, N. B., & Schwartz, J. (2018). The temporal flows of self-tracking: Checking in, moving on, staying hooked. *New Media & Society*, 20(12), 4590–4607.

Lupton, D. (2016a). *The quantified self: A sociology of self-tracking*. Cambridge, UK: Polity Press.

Lupton, D. (2016b). The diverse domains of quantified selves: Self-tracking modes and dataveillance. *Economy and Society*, 45(1), 101–122.

Lupton, D. (2016c). Digital companion species and eating data: Implications for theorising digital data–human assemblages. *Big Data & Society*, 3(1), 2053951715619947.

Lupton, D. (2019). *Data selves: More-than-human perspectives*. Cambridge, UK: Polity Press.

Lupton, D., & Michael, M. (2017). "Depends on who's got the data": Public understandings of personal digital dataveillance. *Surveillance & Society*, 15(2), 254–268.

Lutz, C., & White, G. M. (1986). The anthropology of emotions. *Annual Review of Anthropology*, 15(1), 405–436.

Lutz, C. A. (1988). *Unnatural emotions: Everyday sentiments on a Micronesian atoll & their challenge to Western theory*. Chicago: University of Chicago Press.

Mackenzie, C., & Stoljar, N. (2000). *Relational autonomy: Feminist perspectives on autonomy, agency, and the social self*. Oxford: Oxford University Press.

Mähler, R., & Vonderau, P. (2017). Studying ad targeting with digital methods: The case of Spotify. *Culture Unbound: Journal of Current Cultural Research*, 9(2), 212–221.

Markham, A. (2021). The limits of the imaginary: Challenges to intervening in future speculations of memory, data, and algorithms. *New Media & Society*, 23(2), 382–405.

Martínez, A. G. (2017, November 10). Facebook's not listening through your phone: It doesn't have to. *Wired*. https://www.wired.com/story/facebooks -listening-smartphone-microphone/

Masso, A., & Kasapoglu, T. (2020). Understanding power positions in a new digital landscape: Perceptions of Syrian refugees and data experts on relocation algorithm. *Information, Communication & Society, 23*(8), 1203–1219.

Massumi, B. (2010). The future birth of the affective fact: The political ontology of threat. In M. Gregg & G. Seigworth (Eds.), *The affect theory reader* (pp. 52–70). Durham, NC: Duke University Press.

Maurer, W. M. (2015). Principles of descent and alliance for big data. In T. Boellstorff & B. Maurer (Eds.), *Data, now bigger and better!* (pp. 67–86). Chicago: Prickly Paradigm Press.

Mayer-Schönberger, V., & Cukier, K. (2013). *Big data: A revolution that will transform how we live, work and think*. Boston: Houghton Mifflin Harcourt.

McFall, L., Cochoy, F., & Deville, J. (2017). Introduction: Markets and the arts of attachment. In F. Cochoy, J. Deville, & L. McFall (Eds.), *Markets and the arts of attachment* (pp. 1–21). London: Routledge.

Milan, S., & Treré, E. (2019). Big data from the South(s): Beyond data universalism. *Television & New Media, 20*(4), 319–335.

Miller, D. (1998). *A theory of shopping*. Ithaca, NY: Cornell University Press.

Mol, A. (2008). *The logic of care: Health and the problem of patient choice*. London: Routledge.

Mol, A., & Law, J. (2004). Embodied action, enacted bodies: The example of hypoglycaemia. *Body & Society, 10*(2–3), 43–62.

Nafus, D. (Ed.). (2016). *Quantified: Biosensing technologies in everyday life*. Cambridge, MA: MIT Press.

Nafus, D., & Sherman, J. (2014). Big data, big questions/This one does not go up to 11: The quantified self movement as an alternative big data practice. *International Journal of Communication, 8*, 11.

Neff, G., & Nafus, D. (2016). *Self-tracking*. Cambridge, MA: MIT Press.

Newlands, G. (2021). Lifting the curtain: Strategic visibility of human labour in AI-as-a-Service. *Big Data & Society, 8*(1), 20539517211016026.

Neyland, D. (2019). *The everyday life of an algorithm*. Cham, Switzerland: Palgrave Macmillan.

Ngai, S. (2005). *Ugly feelings*. Cambridge, MA: Harvard University Press.

Nikunen, K. (2015). Politics of irony as the emerging sensibility of the anti-immigrant debate. In R. Andreassen & K. Vitus (Eds.), *Affectivity and race: Studies from Nordic contexts* (pp. 21–41). Farnham: Ashgate.

Noble, S. U. (2018). *Algorithms of oppression: How search engines reinforce racism*. New York: New York University Press.

O'Neil, Cathy. (2016). *Weapons of math destruction: How big data increases inequality and threatens democracy*. New York: Broadway Books.

Pääkkönen, J., Laaksonen, S.-M., & Jauho, M. (2020). Credibility by automation: Expectations of future knowledge production in social media analytics. *Convergence*, *26*(4), 790–807.

Paasonen, S. (2018a). Infrastructures of intimacy. In R. Andreassen, M. Nebeling Petersen, K. Harrison, & T. Raun (Eds.), *Mediated intimacies: Connectivities, relationalities and proximities* (pp. 103–116). London: Routledge.

Paasonen, S. (2018b). Affect, data, manipulation and price in social media. *Distinktion: Journal of Social Theory*, *19*(2), 214–229.

Paasonen, S. (2021). *Dependent, distracted, bored: Affective formations in networked media*. Cambridge, MA: The MIT Press.

Pantti, M., Nelimarkka, M., Nikunen, K., & Titley, G. (2019). The meanings of racism: Public discourses about racism in Finnish news media and online discussion forums. *European Journal of Communication*, *34*(5), 503–519.

Pantzar, M., & Ruckenstein, M. (2017). Living the metrics: Self-tracking and situated objectivity. *Digital Health*, *3*, 2055207617712590.

Pantzar, M., Ruckenstein, M., & Mustonen, V. (2017). Social rhythms of the heart. *Health Sociology Review*, *26*(1), 22–37.

Pariser, E. (2011). *The filter bubble: How the new personalized web is changing what we read and how we think*. New York: Penguin Books.

Pinch, T. J., & Oudshoorn, N. (Eds.). (2005). *How users matter: The co-construction of users and technologies*. Cambridge, MA: The MIT Press.

Pink, S., Berg, M., Lupton, D., & Ruckenstein, M. (Eds.). (2022). *Everyday automation: Experiencing and anticipating emerging technologies*. London: Routledge. https://doi.org/10.4324/9781003170884

Pink, S., & Fors, V. (2017). Being in a mediated world: Self-tracking and the mind–body–environment. *Cultural Geographies*, *24*(3), 375–388.

Pink, S., Lanzeni, D., & Horst, H. (2018). Data anxieties: Finding trust in everyday digital mess. *Big Data & Society*, *5*(1), 2053951718756685.

Pink, S., Ruckenstein, M., Willim, R., & Duque, M. (2018). Broken data: Conceptualising data in an emerging world. *Big Data & Society*, *5*(1), 2053951717753228.

Pink, S., & Salazar, J. F. (2017). Anthropologies and futures: Setting the agenda. In J. F. Salazar et al. (Eds.), *Anthropologies and futures: Researching emerging and uncertain worlds* (pp. 3–22). London: Bloomsbury Academic.

Pitt, S. (2020, August 7). Why we think our phones are secretly listening to us. *Medium*. https://debugger.medium.com/why-we-think-our-phones-are -secretly-listening-to-us-4fd4176a43e3

Pols, J., Willems, D., & Aanestad, M. (2019). Making sense with numbers: Unravelling ethico-psychological subjects in practices of self-quantification. *Sociology of Health & Illness*, *41*(S1), 98–115.

Pöyhtäri, R., Nelimarkka, M., Nikunen, K., Ojala, M., Pantti, M., & Pääkkönen, J. (2019). Refugee debate and networked framing in the hybrid media environment. *International Communication Gazette*, *83*(1), 81–102.

Prainsack, B., & Van Hoyweghen, I. (2020). Shifting solidarities: Personalisation in insurance and medicine. In I. Van Hoyweghen, V. Pulignano, & G. Meyers (Eds.), *Shifting solidarities: Trends and developments in European societies* (pp. 127–151). New York: Palgrave Macmillan.

Pridmore, J., & Lyon, D. (2011). Marketing as surveillance: assembling consumers as brands. In D. Zwick & J. Cayla (Eds.), *Inside marketing: Practices, ideologies, devices* (pp. 115–136). Oxford: Oxford University Press.

Puig de la Bellacasa, M. (2017). *Matters of care: Speculative ethics in more than human worlds*. Minneapolis: University of Minnesota Press.

Rettberg, J. W. (2014). *Seeing ourselves through technology: How we use selfies, blogs and wearable devices to see and shape ourselves*. London: Palgrave Macmillan.

Rettberg, JW. (2018). Apps as companions: How quantified self apps become our audience and our companions. In B. Ajani (Ed.), *Self-tracking: Empirical and philosophical investigations* (pp. 27–42). Basingstoke: Palgrave Macmillan.

Roberts, S. T. (2019). *Behind the screen*. New Haven, CT: Yale University Press.

Rosenfeld, S. A. (2019). *Democracy and truth: A short history*. Philadelphia: University of Pennsylvania Press.

Ruckenstein, M. (2014). Visualized and interacted life: Personal analytics and engagements with data doubles. *Societies, 4*(1), 68–84.

Ruckenstein, M. (2017). Keeping data alive: Talking DTC genetic testing. *Information, Communication & Society, 20*(7), 1024–1039.

Ruckenstein, M., & Granroth, J. (2020). Algorithms, advertising and the intimacy of surveillance. *Journal of Cultural Economy, 13*(1), 12–24.

Ruckenstein, M., & Pantzar, M. (2015). Datafied life: Techno-anthropology as a site for exploration and experimentation. *Techné: Research in Philosophy and Technology, 19*(2), 191–210.

Ruckenstein, M., & Pantzar, M. (2017). Beyond the quantified self: Thematic exploration of a dataistic paradigm. *New Media & Society, 19*(3), 401–418.

Ruckenstein, M., & Schull, N. D. (2017). The datafication of health. *Annual Review of Anthropology, 46*(1), 261–278.

Ruckenstein, M., & Turunen, L. L. M. (2020). Re-humanizing the platform: Content moderators and the logic of care. *New Media & Society, 22*(6), 1026–1042.

Savolainen, L., Trilling, D., & Liotsiou, D. (2020). Delighting and detesting engagement: Emotional politics of junk news. *Social Media & Society, 6*(4), 2056305120972037.

Savolainen, L., & Ruckenstein, M. (2022). Dimensions of autonomy in human–algorithm relations. *New Media & Society*, June 14. https://doi.org /10.1177/14614448221100802.

Schüll, N. D. (2012). *Addiction by design: Machine gambling in Las Vegas*. Princeton, NJ: Princeton University Press.

Schüll, N. D. (2016). Data for life: Wearable technology and the design of self-care. *BioSocieties, 11*(3), 317–333.

Schüll, N. D. (2018, October 31). *The sense mother*. Society for Cultural Anthropology. https://culanth.org/fieldsights/the-sense-mother

Schüll, N. D. (2019). The data-based self: Self-quantification and the data-driven (good)life. *Social Research, 86*(4), 909–930.

Schwartz, S. A., & Mahnke, M. S. (2020). Facebook use as a communicative relation: Exploring the relation between Facebook users and the algorithmic news feed. *Information, Communication & Society, 24*(7), 1–16.

Schwennesen, N. (2019). Algorithmic assemblages of care: Imaginaries, epistemologies and repair work. *Sociology of Health & Illness, 41*(S1), 176–192.

Scott, J. C. (1998). *Seeing like a state: How certain schemes to improve the human condition have failed.* New Haven, CT: Yale University Press.

Seaver, N. (2017). Algorithms as culture: Some tactics for the ethnography of algorithmic system. *Big Data & Society, 4*(2), 2053951717738104.

Seaver, N. (2019a). Knowing algorithms. In J. Vertesi & D. Ribes (Eds.), *DigitalSTS: A field guide for science & technology studies* (pp. 412–422). Princeton, NJ: Princeton University Press.

Seaver, N. (2019b). Captivating algorithms: Recommender systems as traps. *Journal of Material Culture, 24*(4), 421–436.

Sharma, D., & Tygstrup, F. (2015). Introduction. In D. Sharma & F. Tygstrup (Eds.), *Affectivity and the study of culture* (pp. 1–20). Berlin: De Gruyter.

Sharon, T. (2017). Self-tracking for health and the quantified self: Re-articulating autonomy, solidarity, and authenticity in an age of personalized healthcare. *Philosophy & Technology, 30*(1), 93–121.

Sharon, T. (2018). Let's move beyond critique—but please, let's not depoliticize the debate. *American Journal of Bioethics, 18*(2), 20–22.

Siles, I., Segura-Castillo, A., Solís, R., and Sancho, M. (2020). Folk theories of algorithmic recommendations on Spotify: Enacting data assemblages in the global South. *Big Data & Society, 7*(1), 205395172092337.

Skeggs, B. (2020). Algorithms for "hers": In whose interests? *Feminist Media Studies, 20*(5), 733–736.

Skeggs, B., & Yuill, S. (2016). Capital experimentation with person/a formation: How Facebook's monetization refigures the relationship between property, personhood and protest. *Information, Communication & Society, 19*(3), 380–396.

Sloane, M. (2019). Inequality is the name of the game: thoughts on the emerging field of technology, ethics and social justice. In *Proceedings of the Weizenbaum Conference 2019.* https://doi.org/10.34669/wi.cp/2.9

Stark, L. (2018). Algorithmic psychometrics and the scalable subject. *Social Studies of Science, 48*(2), 204–231.

Strathern, M. (Ed.) (2000). *Audit cultures: Anthropological studies in accountability, ethics, and the academy.* London: Routledge.

Suchman, L. A. (2002). Practice-based design of information systems: Notes from the hyperdeveloped world. *The Information Society, 18*(2), 139–144.

Suchman, L. (2007). *Human-machine reconfigurations: Plans and situated actions.* New York: Cambridge University Press.

Tanninen, M., Lehtonen, T.-K., & Ruckenstein, M. (2021). Tracking lives, forging markets. *Journal of Cultural Economy, 14*(4), 449–463.

Tanninen, M., Lehtonen, T.-K., & Ruckenstein, M. (2022a). Trouble with autonomy in behavioral-based insurance. *British Journal of Sociology, 73*(4), 786–798.

Tanninen, M., Lehtonen, T.-K., & Ruckenstein, M. (2022b). The uncertain element: Personal data in behavioural insurance. In K. Booth, S. French, & C. Lucas (Eds.), *Elemental insurance* (pp. 187–200). London: Routledge.

Taylor, C. (1992). Sources of the self: The making of the modern identity. Cambridge, MA: Harvard University Press.

Taylor, L., Floridi, L., & Van der Sloot, B. (Eds.). (2016). *Group privacy: New challenges of data technologies.* Cham, Switzerland: Springer.

Thornham, H. (2019). Algorithmic vulnerabilities and the datalogical: Early motherhood and tracking-as-care regimes. *Convergence, 25*(2), 171–185.

Tierney, T. F. (1993). *The value of convenience: A genealogy of technical culture.* Albany: State University of New York Press.

Tsing, A. (2005). *Friction: An ethnography of global connection.* Princeton, NJ: Princeton University Press.

Turkle, S. (2011). *Alone together: Why we expect more from technology and less from each other.* New York: Basic Books.

Turner, F. (2006). *From counterculture to cyberculture.* Chicago: University of Chicago Press.

Turow, J. (2012). *The daily you: How the new advertising industry is defining your identity and your worth.* New Haven, CT: Yale University Press.

Vaidhyanathan, S. (2012). *The Googlization of everything (and why we should worry).* Berkeley: University of California Press.

Valentine, G. (1989). The geography of women's fear. *Area, 21*(4), 385–390.

Van Dijck, J. (2014). Datafication, dataism and dataveillance: Big data between scientific paradigm and ideology. *Surveillance & Society, 12*(2), 197–208.

Velkova, J., & Kaun, A. (2021). Algorithmic resistance: media practices and the politics of repair. *Information, Communication & Society, 24*(4), 523–540.

Westlund, A. C. (2009). Rethinking relational autonomy. *Hypatia*, 24(4), 26–49.

Wetherell, M. (2013). Affect and discourse—What's the problem? From affect as excess to affective/discursive practice. *Subjectivity*, 6(4), 349–368.

Wiener, N. (1948). *Cybernetics: Or control and communication in the animal and the machine.* Cambridge, MA: MIT Press.

Williams, R. (1961). *The long revolution.* New York: Columbia University Press; London: Chatto & Windus.

Williams, R. (1977). *Marxism and literature.* Oxford: Oxford University Press.

Williamson, B. (2015). Algorithmic skin: Health-tracking technologies, personal analytics and the biopedagogies of digitized health and physical education. *Sport, Education and Society*, 20(1), 133–151.

Winner, L., 1980. Do artifacts have politics? *Daedalus*, 109(1), 121–136.

Wyatt, S. (2004). Danger! Metaphors at work in economics, geophysiology, and the Internet. *Science, Technology, & Human Values*, 29(2), 242–261.

Ylä-Anttila, T. (2020). Social media and the emergence, establishment and transformation of the right-wing populist Finns Party. *Populism*, 3(1), 121–139.

Ytre-Arne, B., & Moe, H. (2020). Folk theories of algorithms: Understanding digital irritation. *Media, Culture & Society*, 43(5), 807–824.

Zelizer, V. (1985). *Pricing the priceless child: The changing social value of children.* New York: Basic Books.

Ziewitz, M. (2016). Governing algorithms: Myth, mess, and methods. *Science, Technology, & Human Values*, 41(1), 3–16.

Ziewitz, M. (2017). A not quite random walk: Experimenting with the ethno-methods of the algorithm. *Big Data & Society*, 4(2), 2053951717738105.

Ziewitz, M. (2019). Rethinking gaming: The ethical work of optimization in web search engines. *Social Studies of Science*, 49(5), 707–731.

Zuboff, S. (2015). Big other: Surveillance capitalism and the prospects of an information civilization. *Journal of Information Technology*, 30(1), 75–89.

Zuboff, S. (2019). *The age of surveillance capitalism: The fight for the future at the new frontier of power.* London: Profile Books.

Zwick, D., & Bradshaw, A. (2016). Biopolitical marketing and social media brand communities. *Theory, Culture & Society*, 33(5), 91–115.

Index

algorithmic systems (*continued*)
participatory models for the
design of, 174–75; predictive
algorithmic systems, 64; problems
of, 169; and promoting feelings of
being entrapped, 173; providers of,
187; spread of, 18–19; study of,
21–22; technical details of, 3–4;
tendency of to ignore or reduce
the participatory efforts of
humans, 186–87

algorithmic techniques, 4, 40, 45, 51,
52, 71, 91–92, 94, 95, 97, 99, 139,
141, 185, 198; pleasures gained
from, 82

algorithmic technologies, 6, 9, 18, 21,
166–67, 199

algorithms: algorithmic interactions,
27; algorithmic literacy programs,
194–95; aspects of a perfect
algorithm, 90–91; association of
with dystopian predictions of the
future, 5–6; changing functions of,
55; as detection mechanisms, 72;
emotionally charged engagements with algorithms, 36–37, 197;
and the feel of algorithms
(including the dominant feel of),
41–42, 50, 57–58, 60–62, 91, 165; as
a flexible cultural object, 33; as
"human too" (human-machine
connections), 3; and friction, 6–10,
133–37, 17; and the interpenetration of technological and human
forces and agencies, 17–18;
mundane experiences with, 21;
neutral feel of, 44; personal
algorithm stories (personal

responses to algorithms), 2–3, 8,
194; pleasures associated with,
69–70, 95; as a "sensitizing
concept," 33; and social research,
20; tyranny of, 85–86; unpleasant
experiences with, 1–2; unsubstantiated comments concerning, 3;
weakness of, 139. *See also*
emotions/feelings; friction, and
algorithms

AlgorithmWatch, 21

Amazon, 7, 16

Andrejevic, Mark, 54, 138

annoyance and irritation (with
advertisements and algorithms),
51, 58–60, 162, 197; articulations
of, 136–37, 142, 148, 149–50;
emerging irritation, 66–68, 139;
sources of, 140–41

anticipation, 24, 99, 124, 134, 143,
163, 186, 190, 200; logic of, 189;
power of, 189

Aoun, Joseph, 75

artificial intelligence (AI), 9, 70;
harmonized rules for in the
European Union, 159; providers
of, 73

automation, 8, 13, 52, 135, 149;
everyday automation, 76–78;
potential of, 125

autonomy, 195; autonomous agency,
48; autonomous vehicles, 74;
balancing algorithmic care with
autonomy, 80–81; loss of, 181;
personal autonomy, 84, 102; and
protection of free will, 182; quest
for, 185–86; tensions with
autonomy, 181–85

dystopias/dystopian futures, 64, 124; shared dystopias, 128–30

emotions/feelings, 31, 45–46, 197; analysis of the structures of feelings, 39–40; articulations of, 32–33, 82–83; attempts to capture feelings, 41; culturally patterned, 24–27; emotional reactions to narrated nuisance or irritation, 24; emotional vocabularies, 28; feelings as agents of history, 40; feminine emotionality, 200; and neutral and pleasurable emotional reactions toward, 34, 44–45; positive feelings about human-algorithm relations, 56–57; shared feelings, 193; structures of, 56–60, 66, 73, 86, 90, 96–97, 100, 102, 110–11, 161, 195, 196; and tracing emotional responses, 25; uncertain nature of, 100; vocabulary of, 38–39
ethics: ethical guidelines and society, 158–59; ethico-political conversation concerning the future, 27–33; feminist ethics, 182
Eubanks, Virginia, 51

Facebook, 50, 51, 65, 86, 104–5, 111, 115, 116, 128, 144; advertising space on, 52; critique of in Finland, 14–15
fear, 100; articulations of, 103; choosing not to fear, 115–18; digital geography of, 100–103, 114, 118, 122, 130–32, 192–93; fear of violence against women, 101–2; gendered geography of, 101;

relocating experiences of from the personal to the collective, 100. *See also* oppositional fear
feedback loops, 16, 17, 20; visibility of, 176–81
feelings. *See* emotions/feelings
Finland, 31, 32, 45, 61, 85, 144, 196; algorithmic culture of, 129; critique of Facebook in, 14–15; data traces in, 144; established history of gender equality in, 138, 170; forward-looking algorithmic culture in Finland, 62; and government fostering of AI, 11–12; as one of the most digitalized countries in the world, 11; public trust in the government of, 12–13; and the study of friction in relation to processes of datafication, 10–11; values of social justice in, 14; violence against women in, 101. *See also* "heating up" (*kuumottaa* [Finnish]); Nordic welfare society
Fors, Vaike, 80
Fourcade, Marion, 45, 47
friction. *See* algorithms, and friction

gender stereotypes, 50; amplification of, 137–39
General Data Protection Regulation (GDPR), 12, 106, 159
Gerlitz, Carolin, 19, 104–5
global living lab, 158–61
Granroth, Julia, 28–29

Healy, Kieran, 45, 47
"heating up" (*kuumottaa* [Finnish]), 39, 121–25; articulations of, 125;

Founded in 1893,
UNIVERSITY OF CALIFORNIA PRESS
publishes bold, progressive books and journals
on topics in the arts, humanities, social sciences,
and natural sciences—with a focus on social
justice issues—that inspire thought and action
among readers worldwide.

The UC PRESS FOUNDATION
raises funds to uphold the press's vital role
as an independent, nonprofit publisher, and
receives philanthropic support from a wide
range of individuals and institutions—and from
committed readers like you. To learn more, visit
ucpress.edu/supportus.